The Missing Element

Andreas Gimsa

Publisher: BoD · Books on Demand GmbH, Überseering 33,

22297 Hamburg, bod@bod.de

Printed by: Libri Plureos GmbH, Friedensallee 273, 22763 Hamburg

ISBN 978-3-6963-8770-9

Price €19.00

The Missing Element

The structure of this publication is based on the following theses, which are discussed, mathematically described, and proven. Where possible, it is explained why the theses correspond with the observations.

Introduction

Why did I write this book? My original goal was to explain the phenomenon of time. In that context, I also considered it essential to integrate the concept of information and the conservation of information into the fundamentals of physics.

The Big Bang theory seemed unnatural to me, for how can everything arise from nothing? Einstein's theory of gravity does not align with my conviction that an electromagnetic explanation of gravity must be possible using Maxwell's equations.

Motivated by this, I formulated new physical equations to describe nature and conducted numerous theoretical and computational experiments. Over many years, these equations were compared with established physics and known experiments through trial and error and further refined. As a result, new physical insights are presented here that provide the key to a new understanding of the world. They have given me deep insights into the nature of the universe and will do the same for the reader. I have endeavored to write the book in such a way that a science student or high school graduate can understand it.

It has long been suspected that there is a so-called ether. To mathematically embed it within physics, the ether is described here within the framework of classical physics using a fifth spacetime dimension, which serves as a general reference frame for all objects contained within it.

The universe is explained as a spherical structure, and the correct spacetime equation is introduced.

It can be shown which physical quantities must change with the age of the universe.

Gravity is described based on an electromagnetic origin derived from Maxwell's equations. Anyone who finds Einstein's spacetime curvature baffling and insufficient as an explanation should read on here.

Readers should be prepared for a few surprises. This publication contains over 15 years evolutionary research in the field of physics. In the physics that describes the universe and gravity, no stone is left unturned. And the spectrum spans from macroscopic physics to quantum physics.

Who knows today that the age of the universe can easily be calculated from elementary quantities and physical constants using Equation (135)? [1]

This is my third science book on the universe. The first book is "The Beauty of Nature" from 2014, the second "The Metric of Spacetime" from 2020, and now "The Missing Element." I have also published many articles. My previous publications contain not only many new and innovative ideas but also mistakes. These are so important to learn from them and ultimately drawing the right conclusions. Today I'm glad I made them, because they helped me move forward. In July 2025, I attended an entrepreneurship day in my hometown. There, a speaker drew attention to a German wordplay. The word "FEHLER" (mistake) contains the letters "HELFER" (helper). That hits the nail on the head—our mistakes are also our helpers. We should engage with them.

With a whopping 333 formulas, this book can certainly be called a formula collection. I read somewhere that many formulas scare readers away and lower book sales. Honestly, I don't care. I have a deep need to share this knowledge with people who don't have to be physicists or scientists.

I welcome constructive discussions and suggestions for improvement.

Note

In numerical representations, English notation is used with respect to periods and commas. Unless other input values are specified for the calculations, I have used the 2022 DODATA values. To ensure clarity, some results are rounded at the end of formulas or in the text.

[1] Andreas Gimsa, The Metric of Spacetime, 2nd edition, Book on Demand GmbH, Hamburg, 2020, ISBN 978-3-00-064784-0, p. 84

1. The universe is spherical; its age is omnipresent

How could one prove that the universe is spherical? Its shape is difficult to measure, as it is vast in size and the distance from our galaxy to the edge of the universe is unknown.

An expansion of the universe is observed that proceeds uniformly when viewed from any location. The following section explains that this observation can be accounted for by a spherical universe whose center remains stationary and whose galaxies move faster the farther they are from the center. This goes so far that a galaxy located directly in front of the outer edge of the universe moves at nearly the speed of light. [2]

How is a sphere in space described mathematically? One way to represent it is the so-called spherical equation, which contains the three spatial coordinates x, y, z and its radius r. This is the set of points in three-dimensional space that are exactly at a distance r from the origin and describe a spherical surface:

$$x^2 + y^2 + z^2 = r^2 = c^2 {t_k}^2 \qquad (001)$$

Is it conceivable that the universe could be described by such an equation? This possibility exists, and it remains to be proven why the following spherical equation (002), with $c_i^2 = -c^2$, can be applied to the universe and the events occurring within it.

In this context, it should also be noted that although the universe has a very low average density (see also Table 2), it satisfies the condition: $R = GM/c^2$ for its radius, under which a black hole exists. This is half of the so-called Schwarzschild radius. The aim is to dispel the myth that black holes are extremely compact gravitational monsters. Figure 1 shows various black holes with masses ranging from $10^{25} kg$ to $10^{53}\ kg$. For a mass in the range of $M = 10^{40}\ kg$, their density is already lower than that of atmospheric air. It is interesting that all black holes possess the same surface-specific entropy. The surface is a measure of potential information: it can contain only as much information as there is space available on the surface.[3] This is a clear indication that all the information of the black hole is stored on its surface.

---

[2] Andreas Gimsa, The Great Fallacy of the Accelerated Expansion of the Universe, International Journal of Scientific Research and Management, Volume 10, Issue 03, March 2022, DOI: 10.18535/ijsrm/v10i3.aa01, p. 100

[3] Brian Greene, The Fabric of the Cosmos, 3rd edition, Goldmann Verlag, Munich, 2008, ISBN 978-3-442-15487-6, p. 537

The entropy of black holes remains constant throughout their evolution. To ensure that no exchange of energy or matter with the environment occurs, thereby maintaining entropy constancy, the radius of a black hole must change at the speed of light. From this, I derive the following interesting propositions: Over time, black holes are constantly growing or shrinking. A black hole has a minimum surface area large enough that all information encoded in its mass can also be stored on its surface. They cannot be heavier than their surface area allows. The information-theoretical description of mass is presented in Chapter 15. There, and in Chapter 10, it is shown that the variable mass of a black hole is derived solely from the evolution of its surface. Since the following constant holds for the volume, surface area A_O, and density ρ of a black hole: $3c^2 = \rho\, GA_O$, its density decreases with increasing surface area, as shown in Diagram 1.

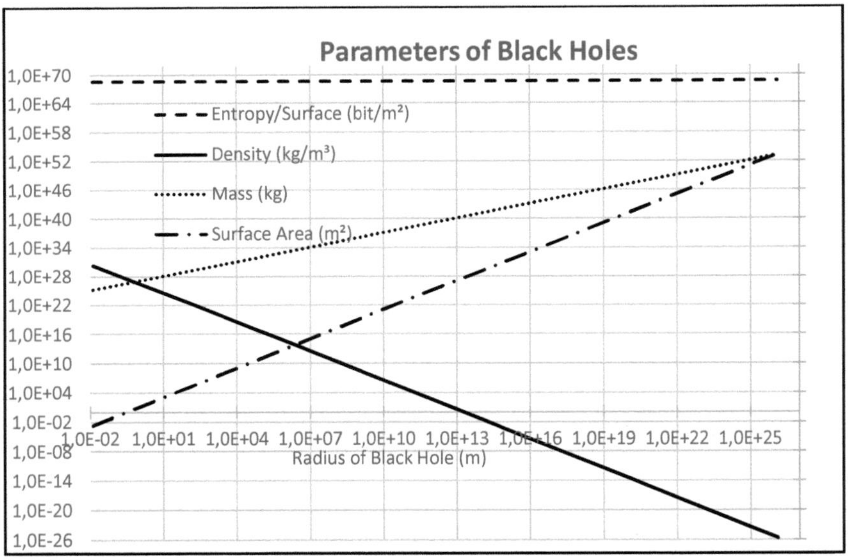

Figure 1: Possible parameters of black holes

Chapter 7 describes the constant entropy of the universe. It is the fundamental prerequisite for the changes in physical constants described in Chapter 13. This allows for a correct description of nature and, ultimately, the unification of quantum physics with gravity.

Why the entropy of the universe does not increase, contrary to current expectations, but remains constant is explained at the end of Chapter 13.

2. The spacetime equation follows from spherical equation of universe

$$\boxed{x^2 + y^2 + z^2 = c^2 t_k^2 = c^2 {t_0}^2 - {c_i}^2 {t_i}^2 = c^2({t_0}^2 + t_i^2)} \tag{002}$$

Equation (002), with its 5 terms, is another way of writing Equation (001). The square of an imaginary number is a negative number, and multiplying a negative number by a negative number yields a positive number. Therefore, the term ${c_i}^2 {t_i}^2$ is positive. The meaning of the imaginary speed of light c_i is explained in Chapter 3. The imaginary time t_i in Equation (002) is interpreted as the present. This must be omnipresent, i.e., the universe must be the same age when viewed from any location.

Equation (003) shows (002) with its 5 terms in differential notation: [4]

$$dx^2 + dy^2 + dz^2 = c^2 d{t_0}^2 - {c_i}^2 d{t_i}^2 \tag{003}$$

I already described imaginary time t_i in connection with imaginary light c_i in the form of ether photons back in 2014.[5] There are already indications of the existence of imaginary time.[6] In general, however, time is still regarded today as real and one-dimensional.

From equations (002) and (003), both a spherical universe and the well-known event cones for causal and non-causal events can be mathematically derived. For this purpose, a complex two-dimensional time was introduced. However, complex time is present in every natural moving system. Since all objects in the universe are in motion, complex time t_k with its real t_0 and imaginary component t_i applies to all objects as well. The magnitude of complex time can be calculated as follows:

$$|t_k| = \sqrt{{t_0}^2 + {t_i}^2} = \sqrt{{t_0}^2 - {t_b}^2} \tag{004}$$

The two components t_0 and t_i of complex time can be interpreted as two temporal dimensions. As described above, the imaginary component t_i is the present. As imaginary time, it is perpendicular to real time.

The mathematical structure of the spherical equation (002) allows for a pulsating universe without an inflationary phase.

The present of the universe t_i can also be defined for arbitrary event intervals.

[4] Andreas Gimsa, The Metric of Spacetime, 2nd edition, Book on Demand GmbH, Hamburg, 2020, ISBN 978-3-00-064784-0, p. 8

[5] Andreas Gimsa, The Beauty of Nature, 2nd edition, Gieselmann Druck und Medienhaus, Potsdam, 2014, ISBN 978-3-923830-97-8, p. 28

[6] Stephen Hawking, A Brief History of Time: From the Big Bang to Black Holes, Bantam Books, New York, 1988, ISBN 978-0-553-10953-5

The time span t_b that elapses while an object is in motion corresponds to the magnitude of the complex time t_k.

As the object's velocity increases, the flow of time on the object is increasingly shifted from the real rest time t_0 into the present t_i. Accordingly, a reduced time span t_b applies. The imaginary time component t_i is not directly measurable, but it can be recognized by the reduced time span t_b at high object velocities. This is demonstrated below in the form of time dilation. For the square of the complex time ${t_k}^2$, the following expression applies:

$$ {t_k}^2 = {t_0}^2 + {t_i}^2 = {t_0}^2 - {t_b}^2 \qquad (005) $$

Equation (005) can be substituted into Equation (002). This yields the spacetime equation (006) with real terms:

$$ x^2 + y^2 + z^2 = c^2{t_k}^2 = c^2({t_0}^2 - {t_b}^2) = {c_i}^2({t_b}^2 - {t_0}^2) $$

$$ \boxed{x^2 + y^2 + z^2 = c^2{t_0}^2 - c^2{t_b}^2} \qquad (006) $$

For the distance x, which can be traversed at the velocity v in the time t_0, the following can be written in one spatial direction using Equation (006):

$$ x^2 = v^2{t_0}^2 = c^2{t_0}^2 - c^2{t_b}^2 \qquad \text{with} \quad y = z = 0 \qquad (007) $$

From this description, the correct formalism of time dilation can be derived, thereby demonstrating that it follows from the space-time equation and not, as is often mistakenly believed, from the Lorentz transformation.

$$ \boxed{v^2{t_0}^2 + c^2{t_b}^2 = c^2{t_0}^2} \qquad (007a) $$

To be on the safe side, the quadratic equation (007) should also be checked using the so-called inverse Pythagorean $\frac{1}{a^2} + \frac{1}{b^2} = \frac{1}{h^2}$ with the height h. In the right-angled triangle described by (007a), the height h can be easily calculated:

$$ h = \frac{v\,t_0\,c\,t_b}{c\,t_0} = v\,t_b \qquad (007b) $$

Now, using the square of the height $v^2\,{t_b}^2$ and the squares of the two legs $v^2{t_0}^2$ and $c^2{t_b}^2$, the inverse Pythagorean theorem holds:

$$ \boxed{\frac{1}{v^2{t_0}^2} + \frac{1}{c^2{t_b}^2} = \frac{1}{v^2{t_b}^2}} \qquad (007c) $$

As with the parallel connection of ohmic resistors, equation (007c) can be rearranged into a different form:

$$ v^2{t_b}^2 = \frac{v^2{t_0}^2\,c^2{t_b}^2}{v^2{t_0}^2 + c^2{t_b}^2} \qquad (007d) $$

Multiplying out equation (007d) yields the following expression:

$$v^2 {t_b}^2 v^2 {t_0}^2 + v^2 {t_b}^2 c^2 {t_b}^2 = v^2 {t_0}^2 \, c^2 {t_b}^2 \qquad (007e)$$

Equation (007e) is simplified, leaving the following expression:

$$v^2 {t_0}^2 + c^2 {t_b}^2 = c^2 {t_0}^2 \qquad (007f)$$

This corresponds again to equation (007a), from which time dilation follows directly. Equation (007c), which is as elegant as equation (007a), can therefore be applied.

What does equation (007c) tell us mathematically? It states that the distance-squared $v^2 {t_b}^2$, which results from the velocity$v$ and the time interval t_b in the moving state of an object, must always be smaller than either of the two distance-squared terms $v^2 {t_0}^2$ and $c^2 {t_b}^2$, which result from the velocity$v$ and the rest time interval t_0, as well as the speed of light c and the time interval t_b. Here, the contraction of the ether with the object is already visible.

3. Time dilation follows from the spacetime equation

With a simple rearrangement, without the Lorentz transformation, equation (007) yields the reduced time interval t_b of the moving system relative to the time interval t_b of its own rest frame: [7]

$$t_b{}^2 = \frac{c^2 t_0{}^2 - v^2 t_0{}^2}{c^2} \quad \text{or} \quad t_b = t_0\sqrt{1 - \frac{v^2}{c^2}} \tag{008}$$

In a two-mass system in which both masses fall toward each other in free fall from an assumed state of rest and reach the final velocities v_1 and v_2 upon contact of their surfaces, the proper time intervals are reduced as follows:

$$t_{1b} = t_{10}\sqrt{1 - \frac{v_1{}^2}{c^2}} \quad \text{(008a)} \qquad \text{and} \qquad t_{2b} = t_{20}\sqrt{1 - \frac{v_2{}^2}{c^2}} \quad \text{(008b)}$$

Equation (008) could thus be derived by introducing complex time into the spacetime equation (002). A Lorentz transformation is not required. It is based on incorrect assumptions, since it ignores the velocity-dependent changes in the masses of the objects relative to their rest states.

Let us examine Equation (005) together with Equation (008). Here, another notation for complex time can be derived:

$$t_k{}^2 = t_0{}^2 - t_b{}^2 = t_0{}^2 - t_0{}^2\left(1 - \frac{v^2}{c^2}\right) = t_0{}^2\left[1 - \left(1 - \frac{v^2}{c^2}\right)\right] = t_0{}^2\frac{v^2}{c^2} \tag{008c}$$

If the object's velocity is $v = 0$, then the complex time is $t_k = 0$. If the object's velocity is theoretically $v = c$, then the complex time becomes $t_k = t_0$.

The shortened time interval t_b of a moving object relative to its state of rest t_0 in Equation (008) <u>cannot</u> be meaningfully applied in a relativistic analysis of two objects in different reference frames, whose roles are interchangeable.

If two objects move uniformly in opposite directions and have a certain relative velocity with respect to each other, it is in principle conceivable to assign the entire velocity to one object and consider the other object to be at rest. It would be equally conceivable to assign a certain portion of the velocity to one object.

---

[7] Andreas Gimsa, The Metric of Spacetime, 2nd edition, Book on Demand GmbH, Hamburg, 2020, ISBN 978-3-00-064784-0, p. 11

The other object then possesses the differential velocity, which results from the relative velocity and the velocity component of the first object.

In a relativistic analysis, however, the principle must always apply that an object in motion has been supplied with energy relative to its own state of rest or relative to another object in the same state of rest. The moving object then possesses a higher-energy state than the same object at rest.

This state is conserved as long as the velocity does not change. Strictly speaking, the relativity of the velocity of two objects relative to each other is freely selectable only for massless objects. However, such objects do not exist in nature.

Let us consider the distance-squared ds^2 in Minkowski space: [8]

$$ds^2 = dx^2 + dy^2 + dz^2 - c^2 {dt_0}^2 \tag{009}$$

or $$s^2 = x^2 + y^2 + z^2 - c^2 {t_0}^2 \tag{009a}$$

If the so-called Lorentz-invariant distance-squared ds^2 is less than zero, this means that two events are causally related and have what is known as a timelike separation. One event can influence the other.

If, however, the distance square ds^2 is greater than zero, this means that two events cannot be causally dependent on each other and possess a so-called spacelike separation. That is, they are so far apart that light could not travel from one event to the other in the time dt_0. [9]

The distance-squared ds^2 from the Minkowski metric corresponds to the term $-c^2 {t_b}^2$ in the spacetime equation from Equation (006). Consequently, this spherical equation can also be used to introduce the event cones of causal and acausal events, just as they can be described using the Minkowski metric.

[8] Hermann Minkowski, Space and Time, Physikalische Zeitschrift, Vol. 10, 1909, pp. 104–111

[9] Walter Greiner, Classical Mechanics I, Wissenschaftlicher Verlag Harri Deutsch, Frankfurt am Main, 2008, ISBN 978-3-8171-1815-1, p. 397

4. Object shortening follows from time dilation

Within an object that is at rest and in rapid motion ($v = 0{,}8\,c$), a light beam travels from point A to point B. In the stationary object, which is one meter long ($l_0 = 1{,}0\,m$), it takes a time t_0 of $3{,}3\,ns$. In the moving object, Equation (008) yields a time that the light beam requires of $2{,}0\,ns$. It necessarily follows that the object's length must contract to $l_b = 0{,}6\,m$. [10]

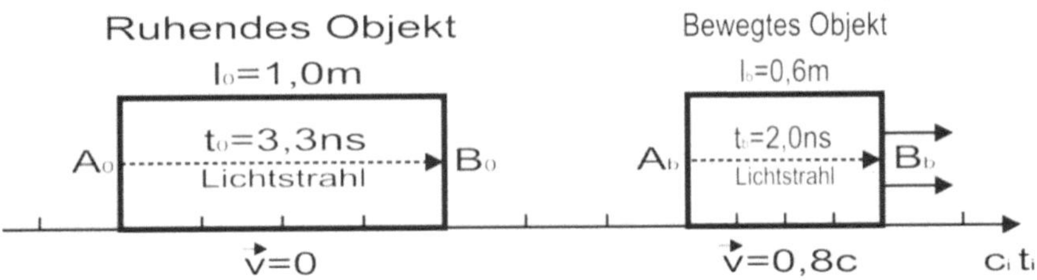

Figure 1: Comparison of the same object at rest and in motion

Therefore, for the length l_b of the moving object, we can write:

$$l_b = c\,t_b = c\,t_0\sqrt{1-\frac{v^2}{c^2}} = l_0\sqrt{1-\frac{v^2}{c^2}} \qquad (010)$$

In a two-mass system in which both masses fall toward each other in free fall from an assumed state of rest and reach the final velocities v_1 and v_2 upon contact of their surfaces, their proper lengths are therefore reduced as follows:

$$l_{1b} = l_{10}\sqrt{1-\frac{{v_1}^2}{c^2}} \quad (010a) \qquad \text{and} \qquad l_{2b} = l_{20}\sqrt{1-\frac{{v_2}^2}{c^2}} \quad (010b)$$

Because the path of light within the object shortens, space (ether) must also contract there.

[10] Andreas Gimsa, The Metric of Spacetime, 2nd edition, Book on Demand GmbH, Hamburg, 2020, ISBN 978-3-00-064784-0, p. 21

5. The ether possesses special physical properties

It has long been known that light in a vacuum travels at a constant speed of $c = 299{,}792{,}458\,m/s$. In Chapter 2, the imaginary speed of light was already introduced via the spacetime equation. [11]

$$c_i = ic \qquad c_i^2 = -c^2 \qquad (011)$$

The imaginary light c_i represents a very essential property of the ether. It makes it possible to define real lengths and masses from the ether.

The virtual photons commonly used to explain quantum processes likely actually exist and represent imaginary light.

Imaginary light can be regarded as the carrier of ordinary light. The magnitude of the total velocity, consisting of that of light and that of the ether, is exactly zero. The two velocities are perpendicular to each other. Ordinary light, for example in the form of a laser beam, can travel in any spatial direction.

In contrast, imaginary light travels in the direction of imaginary time. The direction of imaginary time is perpendicular to the time we know.

$$|v_g| = \sqrt{c^2 + c_i^2} = \sqrt{c^2 - c^2} = 0 \qquad (012)$$

This describes the well-known fact that time stands still for light, which, as an electromagnetic wave, can propagate only in the imaginary ether. While light moves through space at the highest possible speed, time stands still for the photons. They are in the imaginary present t_i. Every object is also permeated by the ether. When an object (see Fig. 1) shortens due to its velocity, the ether within it also contracts. However, it is extremely rigid in this sense and possesses an extremely high modulus of elasticity.

This means that the ether strongly resists compression. It is hypothesized that the imaginary light propagates as a longitudinal wave. That is, the direction of oscillation coincides with the direction of propagation. Therefore, for the imaginary light, propagation in a temporal dimension—the imaginary presentt_i —would be possible.

We will now attempt, theoretically, to accelerate a cube-shaped region of ether to the speed of light. The acceleration of a region of ether is only possible if there is a mass within it.

[11] Andreas Gimsa, The Beauty of Nature, 2nd edition, Gieselmann Druck und Medienhaus, Potsdam, 2014, ISBN 978-3-923830-97-8, p. 23

In the process, the object moves increasingly from real time t into imaginary time t_i. The object contracts with the ether.

As is well known, the propagation speed v_{lw} of longitudinal waves in solids can be calculated using the elastic modulus E_M of the medium and its density ρ.

$$v_{lw} = \sqrt{\frac{E_M}{\rho}} \qquad E_M = \frac{\sigma}{\varepsilon} \qquad (013)$$

Here, σ represents the mechanical stress and ε the strain of the medium.

$$v_{lw} = \sqrt{\frac{\sigma}{\rho\,\varepsilon}} \qquad \rho = \frac{m_b}{V_{co}} \qquad (014)$$

In the system under investigation, the density ρ is the ratio of the relativistically increased mass m_b to its volume V_{co} contracted in the direction of motion.

$$v_{lw} = \sqrt{\frac{\sigma\,V_{co}}{\varepsilon\,m_b}} \qquad \sigma = \frac{F_{co}}{A} \qquad (015)$$

The stressσ is, like pressure, the ratio of a force F to the area A.

$$v_{lw} = \sqrt{\frac{F_{co}\,V_{co}}{\varepsilon\,m_b\,A}} \qquad V_{co} = l^2 l_{co} \quad A = l^2 \qquad (016)$$

The object has a cubic volume V_{co} with a contracted length l_{co}. The area perpendicular to the direction of motion is A.

$$v_{lw} = \sqrt{\frac{F_{co}\,l_{co}}{\varepsilon\,m_b}} \qquad E_{co} = F_{co} l_{co} \qquad (017)$$

The length l_{co} is to be extremely compressed in the direction of motion due to the high velocity: $l_{co} = 0.001 l_0$. The force F_{co}, applied over the compressed distance l_{co}, results in a contraction energy E_{co}. This must be a thousand times greater than the rest energy in order to achieve this compression.

$$v_{lw} = \sqrt{\frac{E}{\varepsilon\,m_b}} \qquad E = m_b c^2 = \frac{m_0 c^2}{\sqrt{1-\frac{v^2}{c^2}}} = E_{co} \qquad (018)$$

If exactly as much contraction energy E_{co} is expended for the compression as the energy equivalent of the mass m_b, it cancels out. A good particle accelerator, for example, could accelerate a proton to a thousand times its rest energy. The proton's diameter would then be compressed to one-thousandth of its initial diameter in the direction of motion.

$$v_{lw} = \sqrt{\frac{c^2}{\varepsilon}} \qquad \varepsilon = \frac{\Delta l}{l_0} \qquad (019)$$

Elongation ε is the ratio of the change in length $\Delta l = l_{co} - l_0$ to the initial length l_0. If the remaining length is approximately $l_{co} \approx 0$, the change in length is $\Delta l \approx -l_0$. The elongation is then $\varepsilon \approx -1$.

This is a nearly complete compression.

$$v_{lw} \approx \sqrt{-c^2} = c_i = ic \tag{020}$$

In the theoretical case considered here, longitudinal ether waves propagate in an extremely compressed cube at nearly the imaginary speed of light c_i. As already mentioned, this is also the speed of the ether.

The rigidity of the ether is not evident from clapping your hands in space. In fact, one would be more likely to conclude that the ether has zero rigidity. Its rigidity is evident in the fact that, as an object's velocity in the direction of space increases, it becomes increasingly difficult to accelerate it further. The faster the object goes, the shorter it becomes, just like the region of ether it occupies. This demonstrates the rigidity of the ether: an extremely large amount of energy is required to achieve even a measurable effect of shortening. One could say that one spacetime dimension is lost in the direction of motion.

As a result, the object moves further and further into imaginary time with this dimension. It is impossible for time to come to a complete standstill on a rapidly moving object according to Equation (008) and for the length to become zero according to Equation (010). The object cannot reach the speed of light; this represents a physical limit. It is determined by the fact that an infinite amount of energy would have to be supplied to an object to achieve this. This is shown by the well-known formula (017) for relativistic energy, because if the object's velocity could become $v = c$ here, the energy required to accelerate the mass would be infinite. That is impossible.

In modern physics, the formula for relativistic velocity addition is frequently invoked to explain why an object cannot reach the speed of light. I reject this, however, because this formula is "proven" using elements of the Lorentz transformation. As already mentioned, the transformation is considered incorrect and worthless because there are no massless inertial systems in nature. Only the object that was previously accelerated experiences an increase in mass, a velocity-dependent time dilation, and a compression.

If there is an ether mass, one must also attribute an energy equivalent E_{ae} to it at the ether velocity c_i:

$$E_{ae} = m_{ae}\, c_i^{\,2} = -m_{ae} c^2 \tag{021}$$

The ether energy equivalent of an ether mass is negative, just like gravitational energy.

In the case of ether photons propagating in the imaginary present t_i with c_i, they could, like photons, be bosons with integer spin.

If one examines the properties of the ether further, it makes sense to look at the properties of the vacuum. For when matter and light are absent, certain properties of the ether remain. It should be noted that the ether permeates everything and is not present only in the vacuum of the universe.

The following known quantities are also among the properties of the ether:

The influence constantε_0, the induction constant $\mu_0 = \frac{1}{c^2 \varepsilon_0}$

the vacuum characteristic impedance $$Z_0 = \frac{p}{e} = \frac{e\,\mu_0}{p\,\varepsilon_0} \qquad (022)$$

with e as the electric elementary charge and p as the magnetic elementary charge. For p, the following applies: $p^2 = 4\pi\,\alpha_{em}\,c\,\mu_0\,\hbar$. [12]

A corresponding relationship is already known for the electric elementary charge. Magnetic elementary charges are discussed in Chapter 14, in the context of gravity.

An isentropic process, such as that occurring during the expansion of the universe, must have an isentropic exponent. Thermodynamics uses the following equation for isentropic processes, with temperature $\mathcal{T}$ and volume V:

$$\mathcal{T}_1 {V_1}^{\kappa-1} = \mathcal{T}_2 {V_2}^{\kappa-1}$$

Furthermore, the expansion proceeds such that as the volume increases linearly, the temperature decreases by the third power. [13]

$$\mathcal{T}_1^{\,3} V_1 = \mathcal{T}_2^{\,3} V_2 \qquad \text{or} \qquad \mathcal{T}_1 {V_1}^{1/3} = \mathcal{T}_2 {V_2}^{1/3} \qquad (023)$$

If one compares the first equation with the second, one sees that the resulting equation is satisfied only if the isentropic exponent:

$$\kappa = {}^{4}/_{3} \qquad (024)$$

Thus, the isentropic exponent of the universe is defined as an essential property of the ether.

---

[12] Andreas Gimsa, The Impedance of Vacuum, Part 1, Kuss GmbH, Potsdam, 2015, ISBN 978-3-00-051924-6, p. 1 ff.

[13] Ralph A. Alpher, Robert Herman, "Remarks on the Evolution of the Expanding Universe," Physical Review 75, 1089–1095, 1949, DOI: 10.1103/PhysRev.75.1089

The ether, via imaginary light traveling at the speed c_i, links together all events occurring at any given location at a specific age of the universe.

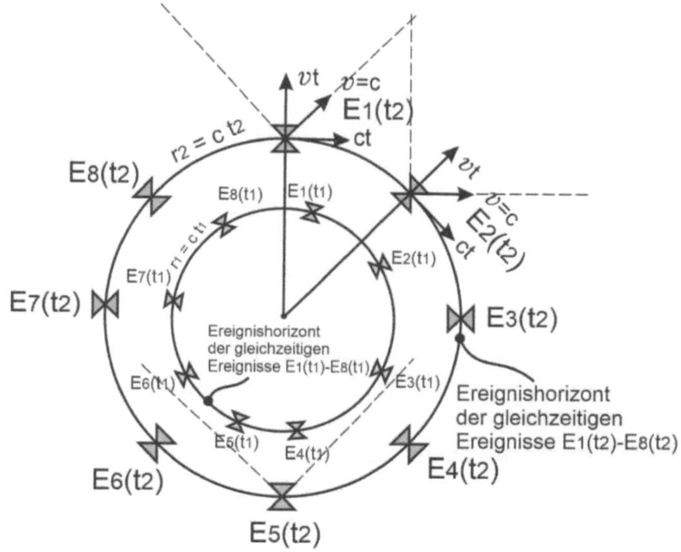

Figure 2: Event horizons of the universe with Minkowski event cones

Figure 2 shows a model cross-section of the expansion of the spherical universe at the two points in time t_1 and t_2. The center of the graph symbolizes the beginning of the expansion; the universe is older toward the outer regions. The double event cones shown indicate the causality of events. As in the Minkowski diagram, events located within the future-facing opening of an event's cone can be influenced by that event. Similarly, events located within the past-facing opening of the event's cone may also have influenced it.

All events at a given point in time lie at different locations on a spherical shell. Such events can only exert an influence on new events at a later point in time (when the universe is older). They are acausal and cannot influence one another.

For example, the cones of the events $E_1(t_2)$ and $E_2(t_2)$ intersect in the distant future. An event that occurs later and lies after this intersection point in time can be influenced by both events.

For example, the event $E_5(t_2)$ may have been influenced by the past events $E_3(t_1)$, $E_4(t_1)$, $E_5(t_1)$ and $E_6(t_1)$. It is quite obvious that an increasing number of past events can influence an event in the future.

So what exactly is the ether? The ether is a universal frame of reference that manifests itself in the form of the omnipresent imaginary present t_i. It corresponds to the respective age t_{uni} of the universe. In this way, viewed from any point in the universe, the respective age of the universe turns out to be identical. The view that the age of the universe represents its present is reasonable simply because today's age of 13.8 billion years is our present. Strictly speaking, tomorrow the age will be 13.8 billion years and one day, and that will be the new present. The imaginary present t_i thus establishes an acausal and instantaneous connection between all simultaneous events in the universe.

Quantum physics describes a vacuum's zero-point energy, and so-called vacuum fluctuations are observed in which quanta arise and annihilate each other if the duration of their existence is sufficiently brief. Quanta can become entangled and instantly transmit their altered state over arbitrary distances via the so-called collapse of the wave function during a state measurement. Thus, such quanta are connected to one another. However, a connection that exists without any kind of medium is mystical and therefore to be rejected from a physical standpoint.

Quantum entanglement can be explained by the fifth dimension of space-time.

With imaginary time, there are consequently 5 dimensions in nature, 3 spatial and 2 temporal: $x, y, z, t \text{ and } t_i$. ***Imaginary time t_i is the missing element in the form of a previously unknown second temporal dimension!*** An imaginary time delay of electromagnetic waves in an absorbing scattering matrix was first systematically investigated experimentally in 2025.[14]

Two events one meter apart cannot be causal if the second occurs exactly 3 nanoseconds after the first. The second event lies within the imaginary time of the first. If, at this spatial distance, the second event were to occur 3.5 nanoseconds after the first, causality would be possible. This is because light takes 3.3 nanoseconds to travel this distance.

For the smallest possible event distance of $1.616\ 10^{-35}\ m$, there is only a single, smallest time span of $5.391\ 10^{-44}\ s$ in which light can traverse it. Two events at this distance could be causal. That is the definition of an instant as the shortest possible interval of real time.

---

[14] Isabella L. Giovannelli, Steven M. Anlage, "Physical Interpretation of Imaginary Time Delay", *Physical Review Letters* 135, 043801 (2025)

However, it is also possible to specify 5 real elements that are equally interrelated and enable the description of nature. These are energy, mass, information, time, and a real length element that incorporates the ether: E, m, I, t, r. *The negative length element* $r = -c_i t_i$ is defined via imaginary time and the imaginary speed of light. *This element thus also represents the ether.*

The ether impulse

An ether momentum p_{ae} can be specified for the motion of a mass in space: What mass and what velocity are referred to in connection with the ether momentum?

$$p_{ae} = m_0\, c_i = m_b\, v_g \qquad (025)$$

The mass m_0 is the rest mass, and the velocity c_i is that of the ether as described above. The mass m_b follows from the moving rest mass m_0, and v_g is the total vector velocity. It results from the magnitude of the vector of the mutually perpendicular motions of the moving mass and the ether (see Fig. 3):

$$v_g = \sqrt{{c_i}^2 + v^2} \qquad (026)$$

Thus, using the stated conservation of momentum, we can write:

$$p_{ae} = m_0\, c_i = m_b \sqrt{{c_i}^2 + v^2} \qquad (027)$$

Rearranging for the moving mass m_b yields the following expression:

$$m_b = \frac{m_0\, c_i}{\sqrt{{c_i}^2 + v^2}} = \frac{m_0\, c}{\sqrt{-1}\sqrt{{c_i}^2 + v^2}} = \frac{m_0\, c}{\sqrt{c^2 - v^2}} \qquad (028)$$

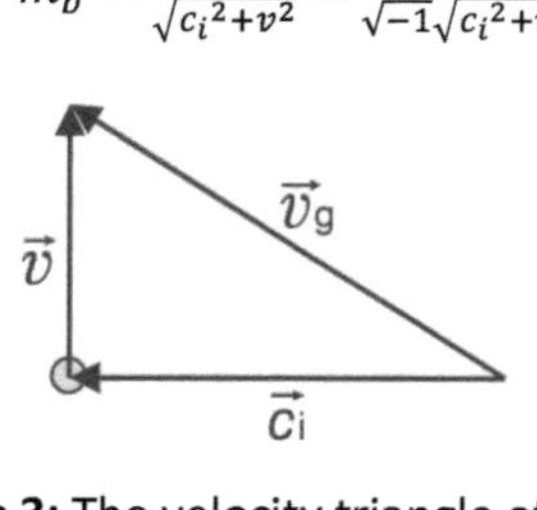

Figure 3: The velocity triangle of the ether momentum

Isn't that a surprise? After the rearrangement, the familiar expression for the moving mass emerges! It follows directly from the conservation of momentum of the mass in the ether.

$$m_b = \frac{m_0}{\sqrt{1 - \frac{v^2}{c^2}}} \qquad (029)$$

In Chapter 10, it will be shown that this expression also follows from the conservation of entropy in gravitation.

6. The great error of the accelerated expansion of the universe

Based on the combination of measured brightness and redshift of standard candles (Type Ia supernovae), it is concluded in connection with Hubble's law that the universe is expanding at an accelerating rate. Here, the brightness indicates how far away a galaxy's supernova is, and the redshift indicates how much the light has been stretched, which serves as a measure of the age of the observed supernova. Hubble's law establishes the relationship between the distance and the velocity of galaxies.

It should be noted that the redshift does not directly measure the age of the universe, but rather the time it took for the light to reach us. It is observed that a more distant galaxy has a higher redshift. To interpret this fact and place it in the context of accelerated expansion, a specific cosmological model is required.

Next, we will describe the spherical structure of a constantly expanding universe, whose center remains stationary and in which galaxies in the outer regions have higher velocities. The velocities of the galaxies remain constant.

The simple formula for the redshift z, is expressed in terms of wavelength or frequency as follows:

$$z = \frac{\lambda_{obs}}{\lambda_0} - 1 = \frac{f_0}{f_{obs}} - 1 \tag{030}$$

Here, λ_{obs} is the observed wavelength and λ_0 is the wavelength in the reference frame. The relativistic velocity-dependent redshift takes the following form when the object and the observer are moving away from each other:

$$z = \sqrt{\frac{c+v}{c-v}} - 1 \tag{031}$$

However, this relationship, which is specifically intended to be relativistic, has so far been rejected for application in cosmology on the grounds that the metric of our spacetime is not a Minkowski metric. It can be shown, however, that Equation (031) holds without restriction.

From the Doppler effect, it is known that the observer's sound frequency f_{obs} decreases as follows with his distance from the source, f_0:

$$f_{obs} = f_0\left(1 - \frac{v}{c}\right) = f_0\left(\frac{c-v}{c}\right) \tag{032}$$

Here, c is the speed of sound in the medium. For electromagnetic waves, the propagation speed is the speed of light. Here, too, a frequency increase occurs as the observer approaches and a reduction as the observer moves away. Whose light frequency is relativistically reduced according to (032), that of the source f_0 or that of the observer f_{obs}? The observer moving away at the speed$v$ has higher energy.

Therefore, in a relativistic analysis with the observer moving at high speed, the following must hold true, just as in relativistic mass increase described later: [15]

$$f_{obs} = \frac{f_0}{\sqrt{1-\frac{v^2}{c^2}}}\frac{c-v}{c} = \frac{f_0}{\sqrt{\frac{c^2-v^2}{c^2}}}\sqrt{\frac{(c-v)^2}{c^2}} = \frac{f_0}{\sqrt{\frac{(c+v)(c-v)}{c^2}}}\sqrt{\frac{(c-v)^2}{c^2}} = f_0\sqrt{\frac{c-v}{c+v}} \qquad (033)$$

If the source is moving away from the observer, the Doppler effect applies:

$$f_{obs} = f_0\frac{1}{1+\frac{v}{c}} = f_0\frac{c}{c+v} \qquad (034)$$

Here, the source has more energy due to its motion at the speed v. Therefore, in a relativistic analysis at high speeds, a relativistic reduction must apply to the observer's frequency f_{obs}:

$$f_{obs} = f_0\sqrt{1-\frac{v^2}{c^2}}\frac{c}{c+v} = f_0\sqrt{\frac{(c+v)(c-v)}{c^2}}\sqrt{\frac{c^2}{(c+v)^2}} = f_0\sqrt{\frac{c-v}{c+v}} \qquad (035)$$

Equations (033) and (035) agree; consequently, it does not matter here which party is moving away.

To be on the safe side, let us also consider the approaching case. As the observer approaches the source, their frequency f_{obs} increases. The well-known Doppler formula applies:

$$f_{obs} = f_0\left(1+\frac{v}{c}\right) = f_0\left(\frac{c+v}{c}\right) \qquad (036)$$

Because the observer is in motion, the relativistic frequency shift f_{obs} must be attributed to him. For his frequency, the following applies from (036):

$$f_{obs} = \frac{f_0}{\sqrt{1-\frac{v^2}{c^2}}}\frac{c+v}{c} = \frac{f_0}{\sqrt{\frac{c^2-v^2}{c^2}}}\sqrt{\frac{(c+v)^2}{c^2}} = \frac{f_0}{\sqrt{\frac{(c+v)(c-v)}{c^2}}}\sqrt{\frac{(c+v)^2}{c^2}} = f_0\sqrt{\frac{c+v}{c-v}} \qquad (037)$$

[15] Andreas Gimsa, The Correct Derivation of the Relativistic Doppler Effect, International Journal of Scientific Research and Management, Volume 10, Issue 06, June 2022, DOI: 10.18535/ijsrm/v10i6.aa01

If, on the other hand, the source approaches the observer, the frequency f_{obs} is given by the Doppler effect:

$$f_{obs} = f_0 \frac{1}{1-\frac{v}{c}} = f_0 \frac{c}{c-v} \tag{038}$$

Now, due to the source's motion, it must become more energetic:

$$f_{obs} = f_0 \sqrt{1 - \frac{v^2}{c^2} \frac{c}{c-v}} = f_0 \sqrt{\frac{(c+v)(c-v)}{c^2}} \sqrt{\frac{c^2}{(c-v)^2}} = f_0 \sqrt{\frac{c+v}{c-v}} \tag{039}$$

Equations (037) and (039) are consistent; consequently, it does not matter here which object is moving toward the other. Everything aligns with observations of nature. And it has absolutely nothing to do with the Special Theory of Relativity, which completely misinterprets relativistic relationships. *What has been demonstrated here is the correct derivation of the relativistic Doppler effect.*

It must be noted that the Doppler frequency shift occurs only as a measurement phenomenon between the source and the observer. Therefore, the expression $c + v$ is also valid, which suggests that there can be a speed exceeding that of light. For example, even with sound, the observer measures a higher frequency as the sound source approaches, even though the source does not possess it.

However, the energy input that leads to a high object velocity increases the object's energy content and results in a genuine increase in the frequency of the radiation it emits.

The universe has a spherical structure; it is in a phase of expansion, and its center is at rest. It is such that a galaxy located further out in it has a higher velocity than a galaxy located further in, i.e., closer to its stationary center.

This will be explained using a model diagram.[16] Fig. 4, A shows the universe at half its current age with smaller objects and distances. Fig. 4, B shows the current universe with the same structure, but with larger objects and distances.

First, the labels in the diagram will be explained.

The point M_P marks the center of the universe $R = 0$; it is at rest $v = 0$. The point R_P marks the radius of the universe $R = R_{uni}$. The point R_P moves outward from the center at the speed of light $v = c$.

[16] Andreas Gimsa, The Great Fallacy of the Accelerated Expansion of the Universe, International Journal of Scientific Research and Management, Volume 10, Issue 03, March 2022, DOI: 10.18535/ijsrm/v10i3.aa01

The galaxies G_2 and G_3 are located at a distance $R = {}^{1}/_{4}\, R_{uni}$ from the center; they are each moving outward at an escape velocity of $v = {}^{1}/_{4}\, c$.

The galaxies G_1 and G_2 , as well as G_2 and G_3 , are separated by a distance of $s_{12} = {}^{1}/_{2}\, R_{uni}$.

The galaxies G_1 and G_3 are separated by a distance of $s_{13} = R_{uni}$. The galaxies G_1 and G_4 are located at a distance of $R = {}^{3}/_{4}\, R_{uni}$ from the center; they are each moving outward at a radial velocity of $v = {}^{3}/_{4}\, c$.

The question to be answered is: at what speeds do the galaxies G_1, G_2, and G_3 move relative to one another? This can be answered using simple velocity addition:

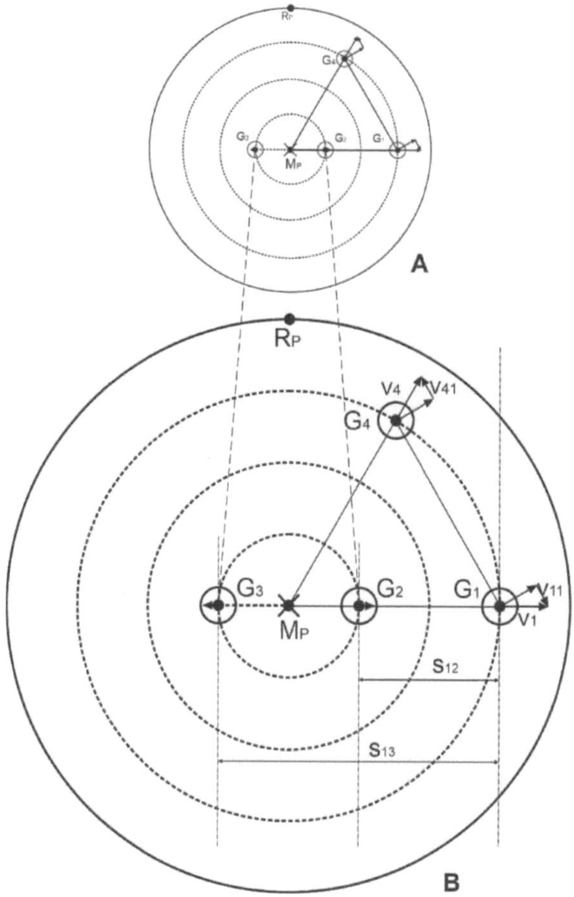

Figure 4: Model of the expanding universe with a stationary center

The galaxy G_1 moves at 3/4 and the galaxy G_2 at 1/4 the speed of light in the same direction. The relative velocity v_{12} corresponds to half the speed of light $v_{12} = 3/4\,c - 1/4\,c = 1/2\,c$.

The galaxy G_1 moves at 3/4 the speed of light and the galaxy G_3 at 1/4 the speed of light in opposite directions. The relative velocity v_{13} corresponds to the speed of light $v_{13} = 3/4\,c + 1/4\,c = c$.

The galaxy G_2 moves at 1/4 the speed of light, and the galaxy G_3 moves at 1/4 the speed of light in opposite directions. The relative velocity v_{23} is half the speed of light $v_{23} = 1/4\,c + 1/4\,c = 1/2\,c$.

The galaxy G_4 has been arranged such that an equilateral triangle is formed on a plane of intersection between it, the center of the universe M_P, and the galaxy G_1. This equilateral triangle remains intact as the universe expands from half its current age to the present (Fig. 4, A to Fig. 4, B). Both galaxies possess the same recession velocity relative to the stationary center of the universe at their respective locations. Does this symmetry—characterized by the equilateral triangle that persists during expansion and the constant distance ratios of all galaxies relative to the radius of the universe—also ensure that the recession velocity between the galaxies G_1 and G_4 is equal to their respective individual recession velocities? This should be verified using velocity triangles. These velocity triangles at the galaxies G_1 and G_4 are clearly visible in Figure 4, B. The radial velocities v_1 and v_4 show the galaxies' outward motion from the center of the universe. The velocities v_{11} and v_{41} are the velocities at which the galaxies move away from each other along their connecting axis. In the right-angled velocity triangle, the following applies to the angle between the radial velocity and the velocity perpendicular to the connecting axis: $\sin \alpha_1 = v_{11}/v_1$ and $\sin \alpha_4 = v_{41}/v_4$. In the equilateral triangle, the angles are $60°$. Because the velocity perpendicular to the connecting axis is thus also perpendicular to one side of the equilateral triangle, the angles α_1 and α_2 must each be $30°$. The sine of 30° is 1/2. It follows that v_1 is twice as large as v_{11} and v_4 is twice as large as v_{41}. This allows us to determine the relative velocity between the galaxies G_1 and G_4:

$$v_{11} + v_{41} = 1/2\,v_1 + 1/2\,v_4 = 1/2\ \ 3/4\,c + 1/2\ \ 3/4\,c = 3/4\,c.$$

This proves that in the universe described here, the escape velocity between these galaxies is as great as their radial escape velocities. All galaxies located on an onion shell of the universe possess a certain radial velocity that increases outward. Isn't that impressive?

Now we must prove that such a universe is also fully consistent with the observed redshifts of the galaxies. To do this, we will use calculations of redshifts derived from various distances from our galaxy (as the observation point) to other galaxies.

Table 1 shows calculations that agree with the observations and verify the universe as depicted in Figure 4.

What velocity is measured using the redshift? Using equations (033) and (035) together with (030), we arrive at the following equation for velocity:

$$v = \frac{cz^2+2cz}{z^2+2z+2} \qquad (040)$$

Table 1 shows the calculation of the redshift at various distances. For high redshift values at great distances from the observed galaxies, the velocity according to (040) approaches the speed of light more and more. But is this a correct measurement?

In the first row, the respective distances of the observed galaxy from ours is determined. In the tenth column of the first row, a distance s_{fix} is chosen that is 99% of the radius of the universe. The second row gives the ratio s/r_{uni} of the distance to the current radius of the universe. The third row shows the velocity of the respective galaxy relative to ours. According to the spherical model of the universe shown in Fig. 4, the velocity, which is proportional to the distance, is given by the ratio s/r_{uni} multiplied by the speed of light. The fourth row specifies the ratio v/c of the galaxy's velocity to the speed of light. It is evident that this ratio agrees with the ratio s/r_{uni} from the second row. The fifth line shows that the ratio $t_{uni} = s/v$ always yields the current age of the universe. In the sixth line, the Hubble constant is calculated using Hubble's law. It is the reciprocal of the age of the universe t_{uni} in SI units. It is calculated precisely for all distances with the respective velocities. In the seventh line, the redshifts$z$ are calculated according to Equation (031). In the eighth line, a check calculation is performed according to Equation (040) for the velocities of the galaxies. The ninth line shows the ratio λ_{obs}/λ_0 of the observed wavelength λ_{obs} to the reference wavelength λ_0 according to Equation (030). It can be seen that at greater distances, the wavelengths λ_{obs} increase, i.e., they are redshifted.

The tenth row shows the respective age of the universe at which the light from the observed galaxy was emitted. The following applies: $t_{start} = t_{uni} - s/c$.

For example, at a short distances , as seen in the first column of the first row, the universe was already nearly as old as it is today at the time the light was emitted.

s (m)	6.54E+22	6.54E+24	1.31E+25	3.27E+25	4.90E+25	6.54E+25	9.81E+25	1.18E+26	1.24E+26	1.29E+26
s/r_{uni}	0.0005	0.0500	0.1000	0.2500	0.3750	0.5000	0.7500	0.9000	0.9500	0.9900
v (m/s)	1.50E+05	1.50E+07	3.00E+07	7.49E+07	1.12E+08	1.50E+08	2.25E+08	2.70E+08	2.85E+08	2.97E+08
v/c	0.0005	0.0500	0.1000	0.2500	0.3750	0.5000	0.7500	0.9000	0.9500	0.9900
t (s)	4.36E+17	4.36E+17	4.36E+17	4.36E+17	4.36E+17	4.36E+17	4.36E+17	4.36E+17	4.36E+17	4.36E+17
Hubble (1/s)	2.29E-18	2.29E-18	2.29E-18	2.29E-18	2.29E-18	2.29E-18	2.29E-18	2.29E-18	2.29E-18	2.29E-18
z	**0.00**	**0.05**	**0.11**	**0.29**	**0.48**	**0.73**	**1.65**	**3.36**	**5.24**	**13.11**
$v_{control}$ (m/s)	1.50E+05	1.50E+07	3.00E+07	7.49E+07	1.12E+08	1.50E+08	2.25E+08	2.70E+08	2.85E+08	2.97E+08
λ_{obs}/λ_0	1.00	1.05	1.11	1.29	1.48	1.73	2.65	4.36	6.24	14.11
t_{start} (10^9a)	**13.8**	**13.1**	**12.4**	**10.4**	**8.6**	**6.9**	**3.5**	**1.4**	**0.7**	**0.1**

Table 1: Redshifts of galaxies at various distances

Figure 2 below shows the relationship between the age of the universe at the time the light was emitted by a galaxy observed at a specific distance and the calculated redshift. The calculation agrees very well with the observations. The shape of the curve in Figure 2 is known in principle and has been determined in this way from measurements of galaxies at specific distances.

The measured cosmological redshifts thus not only indicate the relative velocities between galaxies but, according to the model of the universe shown in Fig. 4, are also a measure of the distances!

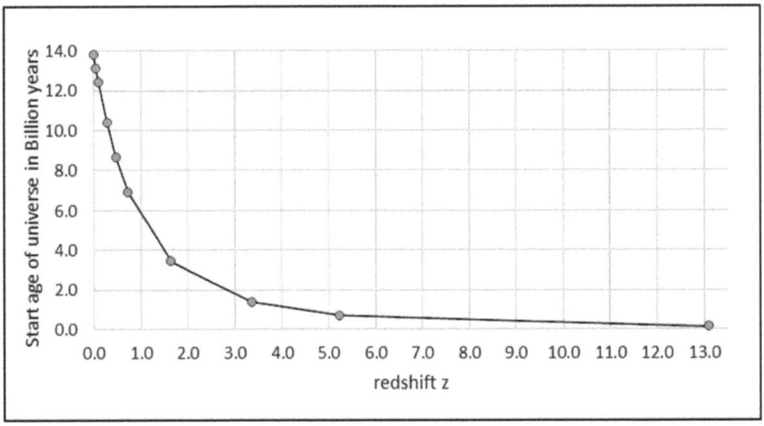

Diagram 2: Age of the Universe at Light Emission and Redshift

Isn't it amazing how everything fits with the observations?

7. The isentropic expansion of the spherical universe

To make the expansion of the universe even clearer, Table 2 shows key physical quantities for four different ages. The time t_s marks the start of expansion, the time t_1 marks the present age, the time t_2 marks one thousand times the present age, and the time t_e marks the end of expansion. All physical quantities are presented in SI units.
The column for the current age also includes the current radius r_{uni} and the current mass m_{uni} of the universe. Using these quantities, a force F_{uni} for the universe can be calculated, which is identical to the Planck force:

$$F_{uni} = G \frac{m_{uni}{}^2}{r_{uni}{}^2} = F_{Planck} = \frac{c^4}{G} = 1.210\ 10^{44}\ N \qquad (041)$$

The equality of the forces F_{uni} and F_{Planck} is fundamental and remains valid even with the change in some "constant of nature" explained in Chapter 12. This astonishing agreement would no longer hold if the radius of the universe were greater than $1.308\ 10^{26}\ m$. With the same density distribution, the mass would have to grow as the cube of the radius of the universe and, with a larger diameter, would also be much higher than what is indicated in the table column for the current age. The agreement according to (041) should therefore be taken into account when it is claimed today that the universe is larger than $1.308\ 10^{26}\ m$.

But when did the expansion of the universe begin? The start time t_s is set in Table 2 as a theoretical 1-Nat information equivalent. The underlying time-information equivalent is derived in Chapter 10. The extent to which the start time $t_{uni\,s}$ may actually have been at the beginning of expansion will be explained. The Planck radius, the Planck mass, and the Planck density are often interpreted as the smallest possible physical quantities at which it might be possible to unify gravity with quantum physics. Table 2 shows that the Planck radius and the Planck mass change with the age of the universe because the gravitational constant G changes (see also Chapter 12). Accordingly, the Planck density, which encompasses both mass and size, also changes. It becomes smaller over time. This means that, when viewed from different ages of the universe, one arrives at different Planck units. It is possible that the Planck time $t_{uni\,s}$ can be regarded as the minimum possible age of the universe.

The minimum energy for storing 1 bit of information, introduced by Brillouin, is calculated as follows : $\Delta E_{1Bit} = k\ \mathcal{T} \ln 2.$ [17]

---

[17] Leon Brillouin, Science and Information Theory, Academic Press Inc., New York, 1956

The minimum energy required to store a natural unit of information of 1 Nat is therefore simply:

$$\Delta E_{1Nat} = k\,\mathcal{T}\ln e = k\,\mathcal{T} \qquad 1\,Nat = 1\,Bit/ln2 \approx 1.44\,Bit \qquad (042)$$

Using the initial temperature of the universe (see Table 2), the mass equivalent for the energy required for 1 Nat is:

$$\Delta m = \frac{k\,\mathcal{T}}{c^2} = \frac{1.381\;10^{-23}\;1.417\;10^{32}}{(2.998\;10^{8})^2} kg = 2.176\;\;10^{-8}\;kg \qquad (043)$$

This yields the well-known Planck mass. The following applies to the change in the mass of the universe over time (see also Chapter 13):

$$\frac{\Delta m_{uni}}{m_{uni}} = -\frac{\Delta t_{uni}}{t_{uni}} \qquad \Delta t_{uni} = -\frac{\Delta m_{uni}}{m_{uni}}\,t_{uni} \qquad (044)$$

As time passes, i.e., as it increases, the mass decreases. This explains the negative sign. Let's examine the time interval from Equation (044) using current values:

$$|\Delta t_{uni}| = \frac{2.176\;10^{-8}}{1.761\;10^{53}}\;4.362\;\;10^{17}\;s = 5.391\;\;10^{-44}\;s \qquad (045)$$

The time obtained there is the Planck time at the start of the universe's expansion (see Table 2). It is thus the smallest time interval.

The table rows will be briefly explained. The numerical values in Table 2 can be verified using the calculation rules provided there.

<u>Line 1</u>: The minimum age of the universe as the starting time $t_{uni\,s}$ assumes that there must be a minimum age for the beginning of the expansion. This is supposed to be the Planck time. The current age $t_{uni\,1}$ of 13.8 billion years is considered established.

Using an information-theoretical basis for the universe, it is possible to determine how long the expansion will last. At the end of the expansion $t_{uni\,e}$, the maximum possible address length must be available (see line 3).

The time $t_{uni\,s}$ at the start of expansion must be multiplied by the number of address Nats (see Chapter 9, p. 37) in the 24th row of Table 2 (for entropy, see also row 18). This yields the age $t_{uni\,e}$ at the end point of expansion.

<u>Line 2</u>: The formula for the current information of the universe is derived in Chapter 10. It is thus directly proportional to time and a measure of the address length of the universe's information (see row 3).

<u>Line 3</u>: The current information of the universe is the number of Nats required to address a microstate.

This number of Nats corresponds to the address length. It is calculated from the current information in line 2, divided by the magnetic diffusivity from Chapter 10. The address length encodes the address space of the microstates.

Line 4: The radius of the universe sphere is calculated by multiplying the respective age by the speed of light.

Line 5: The surface area of the universe sphere can also be calculated using its radius. It is a measure of the elapsed time in the universe.

Line 6: The universe has a spherical volume, which is also derived from the radius (see also Chapter 1 and Chapter 6).

Line 7: Here, we deliberately speak not of a gravitational constant, but of a gravitational value that is not constant. It increases quadratically with the age of the universe (see also Chapter 13).

Line 8: The mass of the universe must decrease with its age (see also Chapter 13).

Line 9: Einstein's energy-mass equivalence applies. As the universe expands, energy decreases proportionally with mass.

Line 10: The density of the universe undergoes a massive decrease from the start of the expansion to its end. The calculation takes into account an average spherical density.

Line 11: The evolution of the temperature of the background radiation is calculated; this behaves proportionally to mass, meaning it decreases.

Line 12: This temperature $\mathcal{T}_{e\,uni}$, referred to here as the ether temperature—which is significantly colder than the background radiation and extremely small—can be determined directly from the age of the universe and could be the radiation temperature of its pulsation.

Line 13: This extremely small pulsation frequency can be calculated using the total expansion time, given that the period of a vibration of the universe consists of one half expansion phase and the other half contraction phase.

Line 14: The "gas constant" of the universe is variable; it increases as the ether temperature decreases according to Line 12, which becomes smaller with age.

Line 15: The isobaric heat capacity of the universe increases as the "gas constant" increases, as stated in line 14.

Line 16: The isochoric heat capacity of the universe also increases with the increase in the "gas constant" as per line 14.

Line 17: The average pressure of the universe decreases dramatically with its age. This is due in particular to the volume increasing as the cube of the radius.

Line 18: The extremely interesting insight that entropy, which remains constant with age, represents the potential information of the universe, is explained in Chapter 9 under the heading of conservation of information. The isentropic expansion of the universe was already discussed in Chapter 5 as a property of the ether.

Line 19: The Planck mass is a known quantity. The only thing that remains unknown is that it must change with the age of the universe; it becomes smaller.

Line 20: Planck time is also a known quantity. It changes proportionally with the age of the universe; it increases.

Line 21: The Planck radius, as a known quantity, changes proportionally with the age of the universe just like Planck time; it also increases.

Line 22: The Planck area is also known and is calculated like the surface area of a sphere with the Planck radius. The Planck area is often regarded as the area of a square. It increases quadratically with time.

Line 23: The known Planck temperature has the same calculation formula as the ether temperature according to line 12. The difference lies solely in the time specification. While the age of the universe is factored into the ether temperature, the Planck temperature is calculated using Planck time. It decreases with age.

Line 24: The entropy in Nats can be derived from line 18 for thermodynamic entropy. In a universe grounded in information theory, it is, so to speak, the measure of all things. It addresses the maximum possible number of microstates of the system and thus represents the macroscopic state of the universe. Line 18 shows that this state can be described by its mass and age. This constant entropy in Nats from line 24 is incorporated into the calculation of the age $t_{uni\,e}$ at the end of the expansion in line 1.

It is interesting to note that the ratio of the mass of the universe to the Planck mass, as well as the ratio of the age of the universe to the Planck time, remains constant during the expansion. This means that the Planck units, which are calculated using the Newtonian gravitational constant, must change over time.

$$m_{uni}/m_{Planck} = t_{uni}/t_{Planck} = \sqrt{S_{Nat\,uni}} = \sqrt{6.546\ 10^{121}} = 8.091\ 10^{60}.$$

Nr.	Physical Quantity	Formula	Time $t_{uni\,s}$	Time $t_{uni\,1}$	Time $t_{uni\,2}$	Time $t_{uni\,e}$
1	Age Universe [s]	***$t_{uni\,1}$ = 13.8 E+09 years***	***5.391E-44***	**4.362E+17**	4.362E+20	3.529E+78
2	Information [m²/s]	$I_{1\,MZ\,uni} = 4\pi\, t_{uni} c^2$	6.089E-26	**4.926E+35**	4.926E+38	3.986E+96
3	Information [Nat]	$I_{Nat\,1\,MZ\,uni} = \frac{4\pi\, t_{uni}\, c^2}{4\pi\, t_{uni\,s}\, c^2}$	1.000E+00	**8.091E+60**	8.091E+63	6.546E+121
4	Radius [m]	$r_{uni} = c\, t_{uni}$	1.616E-35	**1.308E+26**	1.308E+29	1.058E+87
5	Surface Universe [m²]	$A_{uni} = 4\pi\, r_{uni}^{\;2}$	3.283E-69	**2.149E+53**	2.149E+59	1.407E+175
6	Spherical Volume [m³]	$V_{uni} = \frac{4}{3}\pi\, r_{uni}^{\;3}$	1.769E-104	**9.366E+78**	9.366E+87	4.960E+261
7	Gravitational Value [m³ /(kg s²)]	$G_{uni\,x} = \frac{t_{uni\,x}^{\;2}}{t_{uni}^{\;2}} G_{uni}$	1.020E-132	**6.674E-11**	6.674E-05	4.369E+111
8	Mass [kg]	$m_{uni} = \frac{t_{uni}\, c^3}{G}$	1.425E+114	**1.761E+53**	1.761E+50	2.176E-08
9	Energy [J]	$E_{uni} = m_{uni}\, c^2 = p_{uni}\, V_{uni}$	1.280E+131	**1.583E+70**	1.583E+67	1.956E+09
10	Density [kg/m³]	$\rho_{uni} = \frac{m_{uni}}{V_{uni}}$	8.055E+217	**1.880E-26**	1.880E-38	4.388E-270
11	Temperature BG Universe [K]	$\mathcal{T}_{uni\,x} = \frac{m_{uni\,x}}{m_{uni\,1}} \mathcal{T}_{uni\,1}$	2.205E+61	**2.725E+00**	2.725E-03	3.368E-61
12	Temperature Ether Universe [K]	$\mathcal{T}_{e\,uni} = \frac{\hbar}{k_B\, t_{uni}}$	1.417E+32	**1.751E-29**	1.751E-32	2.164E-90
13	Pulsation frequency Universe [Hz]	$f_{uni} = \frac{1}{T_{uni}} = \frac{1}{2(t_{uni\,e} - t_{uni\,s})}$	1.417E-79	**1.417E-79**	1.417E-79	1.417E-79
14	Gas Constant [J/(kg K)]	$R_{uni} = \frac{E_{uni}}{m_{uni}\, \mathcal{T}_{e\,uni}}$	6.344E-16	**5.132E+45**	5.132E+48	4.152E+106
15	Isobaric Heat Capacity [J/(kg K)]	$C_{p\,uni} = \frac{R_{uni}}{1 - 1/\kappa}$	2.537E-15	**2.053E+46**	2.053E+49	1.661E+107
16	Isochoric Heat Capacity [J/(kg K)]	$C_{V\,uni} = C_{p\,uni} - R_{uni}$	1.903E-15	**1.540E+46**	1.540E+49	1.246E+107
17	Pressure [Pa]	$p_{uni} = \frac{m_{uni}\, R_{uni}\, \mathcal{T}_{e\,uni}}{V_{uni}}$	7.240E+234	**1.690E-09**	1.690E-21	3.943E-253
18	Entropy [J/K]	$S_{uni} = \frac{k_B}{\hbar} m_{uni}\, t_{uni}\, c^2 = \frac{E_{uni}}{\mathcal{T}_{e\,uni}}$	9.037E+98	**9.037E+98**	9.037E+98	9.037E+98
19	Mass Planck [kg]	$m_{Planck} = \sqrt{\hbar\, c/G}$	1.761E+53	**2.176E-08**	2.176E-11	2.690E-69
20	Time Planck [s]	$t_{Planck} = \sqrt{\hbar\, G/c^5}$	6.664E-105	**5.391E-44**	5.391E-41	4.362E+17
21	Radius Planck [m]	$r_{Planck} = c\, t_{Planck}$	1.998E-96	**1.616E-35**	1.616E-32	1.308E+26
22	Surface Planck [m²]	$A_{Planck} = 4\pi\, r_{Planck}^{\;2}$	5.015E-191	**3.283E-69**	3.283E-63	2.149E+53
23	Temperature Planck [K]	$\mathcal{T}_{Planck} = \frac{\hbar}{k_B\, t_{Planck}}$	1.146E+93	**1.417E+32**	1.417E+29	1.751E-29
24	Entropy [Nat]	$S_{Nat\,uni} = \frac{S_{uni}}{k_B} = \frac{A_{uni}}{A_{Pl}}$	6.546E+121	**6.546E+121**	6.546E+121	6.546E+121

Table 2: Physical quantities in the expansion of the universe

The cosmic microwave background and the number of stars in the universe [18]

The background radiation could be stellar radiation reflected from the event horizon in a pulsating universe. Since space is isentropic and its radius expands at the speed of light, the radiation cannot escape. It forms the typical profile of a blackbody in spherical space. The Stefan-Boltzmann law then directly leads to the size of the radius of a single protostar, whose surface area is as large as the current sum of the surface areas of all stars at their average temperature. Theoretically, it emits just as much radiation as all the stars that exist today.

$$r_{uni}^{2}\,\mathcal{T}_{bg}^{\;4} = r_{bigstar}^{\;2}\,\mathcal{T}_{star}^{\;4} = const. \tag{046}$$

This directly implies the theoretical radius of a giant star whose surface area is as large as the combined surface areas of all stars today:

$$r_{bigstar} = \frac{\mathcal{T}_{bg}^{\;2}}{\mathcal{T}_{star}^{\;2}} r_{uni} = \frac{2{,}725^2}{5{,}800^2} 1.3\ 10^{26}\ m = 2.88\ 10^{19}\ m \tag{047}$$

If we now calculate its radiative surface area and then compare it to that of a typical star, we also arrive at the approximate number of stars in the universe. This typical star is to be our Sun:

$$N_{star} = \frac{A_{bigstar}}{A_{star}} = \frac{r_{bigstar}^{\;2}}{r_{star}^{\;2}} = \left(\frac{2.9\ 10^{19}}{7.0\ 10^{8}}\right)^2 = 1.72\ 10^{21}\ Sterne \tag{048}$$

What is the current estimate of the number of stars in the universe? As is typical in science, there are various figures on this subject. The main focus is on the order of magnitude. For example, Börner[19] estimates the number of stars in our Milky Way at 100 billion. Based on observations made with the Hubble Space Telescope, the number of galaxies is estimated at 10 billion.

$$100\ 10^9\ \frac{Sterne}{Galaxie}\ x\ 10\ 10^9\ Galaxien = 1.0\ 10^{21}\ Sterne \tag{049}$$

In terms of magnitude, the number of stars calculated using Equation (049) agrees with the number counted from observations. Stellar radiation could therefore actually be in radiative equilibrium with the horizon of space and lead to background radiation.

[18] Andreas Gimsa, The Metric of Spacetime, 2nd edition, Book on Demand GmbH, Hamburg, 2020, ISBN 978-3-00-064784-0, p. 73

[19] Gerhard Börner, Cosmology, Fischer Taschenbuch Verlag, Frankfurt am Main 2011, ISBN 978-3-596-19023-2, pp. 4–6

8. The Pulsation of the Spherical Universe

I have long held the view that the universe must pulsate.[20] If everything in the world vibrates, why shouldn't the universe, of all things, do the same? The idea that everything arose from nothing in a Big Bang is, in my opinion, completely out of touch with reality. Chapter 9 discusses the conservation of potential information in the universe, which I have been documenting in my publications for many years. This important principle is also ignored by the untenable Big Bang theory.

What might the pulsation of the universe look like?

A one-dimensional wave equation, such as that applicable to a guitar string, can be derived with the aid of a few fundamental equations.

For the change in force over the distance segment ∂x of the string, the following applies:

$$\frac{\partial F}{\partial x} = \frac{F_0}{r} \tag{050}$$

Here, the coefficient $1/r$ represents the curvature of the string. The change in force on the plucked string produces an acceleration of the mass in section ∂x:

$$a = \frac{\partial^2 y}{\partial t^2} = \frac{\partial F}{m} = \frac{\partial F}{\rho\, A\, \partial x} \tag{051}$$

Using Equation (050), the acceleration is given by:

$$\frac{\partial^2 y}{\partial t^2} = \frac{1}{r}\frac{F_0\, \partial x}{\rho\, A\, \partial x} = \frac{1}{r}\frac{F_0}{\rho\, A} \tag{052}$$

The expression F_0/A is the string tension in the initial state. From this, the restoring force of the oscillating system after excitation is derived:

$$\sigma = \frac{F_0}{A} \tag{053}$$

The curvature $1/r$ of the string is the second derivative, defined as the slope of the function value of the oscillation with respect to displacement:

$$\frac{1}{r} = \frac{\partial^2 y}{\partial x^2} \tag{054}$$

Using equations (052) and (053), equation (054) yields the well-known one-dimensional wave equation:

$$\frac{\partial^2 y}{\partial x^2} - \frac{\rho}{\sigma}\frac{\partial^2 y}{\partial t^2} = 0 \tag{055}$$

[20] Andreas Gimsa, The Metric of Spacetime, 2nd edition, Book on Demand GmbH, Hamburg, 2020, ISBN 978-3-00-064784-0, p. 64

Equation (055) can be immediately transformed into a spherical wave equation for a spatial wave with the spatially and temporally varying radius r:

$$\frac{\partial^2 r}{\partial x^2} + \frac{\partial^2 r}{\partial y^2} + \frac{\partial^2 r}{\partial z^2} - \frac{\rho}{\sigma}\frac{\partial^2 r}{\partial t^2} = 0 \qquad (056)$$

Equivalent to the spacetime equation with complex time t_k, the differential equation for the pulsation of the spherical universe can now be stated as follows:

$$\frac{\partial^2 r}{\partial x^2} + \frac{\partial^2 r}{\partial y^2} + \frac{\partial^2 r}{\partial z^2} = \frac{3}{c^2}\frac{\partial^2 r}{\partial {t_k}^2} \qquad (057)$$

Here, the ratio is $\rho/\sigma = 3/c^2$. The right-hand term, as the second derivative of the function value of the wave equation with respect to time, is thus the curvature of the surface of the complex sphere—which varies due to the pulsation of space—at the respective age (cf. also (054) and the associated unit: 1/m).

9. The Conservation of Information in the Universe

The conservation of entropy in the universe is a fundamental law of nature. It is as fundamental as the conservation of energy and incorporates time (Equation (60)). [21]

As already mentioned in Chapter 07, time encodes the current address length of the universe. This information requires a universal storage medium. It is conceivable that such a storage medium exists in the form of the surface of the spherical universe.

Similar to the expression for temperature known from Wick rotation, where $\mathcal{T} = \hbar/k\,\tau$ and τ represent the length of the imaginary time axis, a specific temperature for the universe can be determined. As explained in Chapter 5, the present of the universe, t_i, is imaginary as its age. In Wick rotation, imaginary time is understood as the length of the imaginary time axis and is treated as a real value.

If the age of the universe is to be described using a temperature, a slightly modified expression is used, which is explained in Chapter 10. This temperature is identified there as time radiation.

$$\mathcal{T}_{uni} = \frac{\hbar}{k\, t_{uni}} \tag{058}$$

The age of the universe t_{uni} can be interpreted, in the sense of Wick rotation, as the length of its imaginary time axis. The specific temperature given by equation (058) is extremely low given the current age of the universe of 13.8 billion years. It is $1.751\ 10^{-29}$ K. This temperature should not be confused with the cosmic microwave background radiation. The entropy of the universe can now be simply expressed as:

$$S_{uni} = \frac{E_{uni}}{\mathcal{T}} = \frac{m_{uni}\, c^2}{\mathcal{T}} = \frac{r_{uni}\, c^4}{\mathrm{G}\,\mathcal{T}} = \frac{c\, t_{uni}\, c^4}{\mathrm{G}\,\mathcal{T}} = \frac{t_{uni}\, c^5}{\mathrm{G}\,\mathcal{T}} \tag{059}$$

The mass of the universe is currently $1.761\ 10^{53}$ kg and its entropy $9.037\ 10^{98}$ J/K. Using equations (058) and (059), the following concise expression for the entropy of the universe is obtained, which deserves to be called *the world formula*:

$$S_{uni} = \frac{k}{\hbar}\, m_{uni}\, t_{uni}\, c^2 = const. \qquad m_{uni}\, t_{uni} = const. \tag{060}$$

Equation (060) for the entropy of the universe yields the same result as (059). Equation (060) can also be written as the number of N address-Nats:

[21] Andreas Gimsa, The Beauty of Nature, 2nd edition, Gieselmann Druck und Medienhaus, Potsdam, 2014, ISBN 978-3-923830-97-8, p. 42

$$N = S_{Nat\,uni} = \frac{S_{uni}}{k} = \frac{m_{uni}\,t_{uni}\,c^2}{\hbar} = \frac{m_{uni}\,r_{uni}\,\mathrm{c}}{\hbar} = 6.546\;10^{121} \qquad (061)$$

This number N, is the number of Nats required to store a microstate of the universe. At the end of the universe's expansion, only one microstate with N Nats is realized from the vast selection according to (063).

Ludwig Boltzmann's well-known formula for entropy is:

$$S_{uni} = k \ln W \qquad (062)$$

Here, W is the thermodynamic probability that specifies the possible number of microstates of the universe.

$$W = e^{S/K} = e^{N} = e^{6.546\;10^{121}} \qquad (063)$$

This is an incredibly large number with an exponent of 121 zeros. So it is not the number e itself that has 121 zeros, but the exponent of e.

How many Nats are needed to store all microstates?

$$N_{Nat\,\Sigma\,MZ} = W\,N = e^{6.546\;10^{121}}\;6.546\;10^{121} \qquad (064)$$

This number is also of incomparable magnitude and exceeds the number of microstates from (063) by a factor of $6.546\;10^{121}$.

According to equation (060), the product of the mass m_{uni} and the age t_{uni} of the universe is constant. As the age increases, the mass decreases. This means that the universe, along with all its stars and other masses, becomes lighter over time. This can be seen as a violation of the law of conservation of energy, which assumes that energy cannot be lost but can only be converted from one form to another. However, it is hypothesized here that not only does the mass energy decrease, but the negative gravitational energy also increases to the same extent, meaning its negative value becomes smaller over time.

$$E_{mass\,uni} + E_{grav\,uni} = 0 \qquad m_{uni}c^2 - \frac{G{m_{uni}}^2}{r_{uni}} = 0 \qquad (065)$$

Let us again consider the universe as a sphere. If it has a homogeneous density distribution, the gravitational binding energy is known to be only $E_{grav\,uni} = -\frac{3}{5}\frac{G{m_{uni}}^2}{r_{uni}}$. This is not sufficient to explain the total mass loss according to Equation (065). If the density is concentrated at the center of the sphere, the gravitational energy increases. It is assumed that the density increases as the radius of the sphere decreases toward its center in the ratio $\rho{\sim}1/r$ and that there is a finite density at the center of the sphere in the form of the Planck density as the maximum limiting density, which occurs there only in the range of a Planck length.

The result of the integration yields the gravitational binding energy according to the relationship given by Equation (065): $E_{grav\,uni} = -\frac{Gm_{uni}{}^2}{r_{uni}}$.

However, the magnitudes of both forms of energy—mass energy and gravitational energy—decrease over time. If Equation (065) holds, an energy equilibrium exists. The slight decrease in the mass of the universe over small time intervals is not easily measurable, yet it is fundamental to our understanding of the universe.

One could argue as follows: Since the energy equilibrium according to Equation (065) holds, the density profile of the universe must also be such that the desired relationship for gravitational energy is obtained.

It should also be noted that in Equation (065), not only does mass decrease with the age of the universe, but the radius also increases. For the Newtonian gravitational constant G, this means that it must grow quadratically with time.[22] The necessary modification of various supposed natural constants will be discussed in Chapter 13.

Equation (060) shows that energy must be expended to create new time when updating information.

Can formula (064) for the information content of the spherical universe be verified in any way? That is certainly possible. According to Bekenstein, the maximum information content of a system is proportional to its surface area and not to the volume it contains. This is consistent with the discussion of black holes in Chapter 1.

Let us assume that the surface of the spherical universe is composed at all times of surfaces given by Planck surfaces. This allows us to immediately calculate the number N of Planck surfaces:

$$N = \frac{A_{O\,uni}}{A_{O\,Planck}} = \frac{4\,\pi\,r_{uni}{}^2}{4\,\pi\,r_{Planck}{}^2} = \frac{c^2 t_{uni}{}^2}{\frac{\hbar\,G}{c^3}} = \frac{c^5\,t_{uni}{}^2}{\hbar\,G} = 6.546\;10^{121} = const. \quad (066)$$

Since both the surface of the universe and the Planck surface change over time, Equation (066) holds at all times.

Equation (066) yields the number of Planck surfaces N, which simply corresponds to the value S_{uni}/k. Once again, we obtain the address length of the universe in Nats for the encoding of one of its microstates. The value N is the number of address Nats (see also Chapter 15, p. 68).

[22] Andreas Gimsa, The Beauty of Nature, 2nd edition, Gieselmann Druck und Medienhaus, Potsdam, 2014, ISBN 978-3-923830-97-8, p. 75

Formula (060) also applies to gravity and has been verified using it. In the case of gravity as well, the product of mass and time remains constant. [23]

If two masses attract each other from their respective rest positions at a certain distance, their velocities increase until they collide. In the process, their masses increase due to their velocities (from m_0 to m_b respectively). Time intervals become shorter due to their velocities (from t_0 to t_b respectively). Thus, time is reduced during the gravitational process. A shorter time interval means that time passes more slowly. Compared to the flow of time at rest, it is reduced at higher speeds. This also means that fewer events can occur on the same object than on the same object at rest. Thus, less happens on the fast-moving object. Because mass increases with speed, it can also be said that the flow of time decreases with increasing mass or its energy content. In his Special Theory of Relativity, Einstein establishes geometric relationships between two objects moving relative to one another while disregarding the objects' masses. This is fundamentally incorrect and leads to the conclusion, based on his theory, that objects with a specific rest mass are treated exactly the same as objects that are in motion and possess a mass greater than their rest mass. Within the framework of his theory, Einstein considers point-like, massless objects that do not, however, exist in nature. Therefore, this theory is not only worthless, but it has even caused great confusion over many years with the help of its dogmatic proponents. It was drummed into people's minds against logic and intuition. With the falsification of this theory, the unification of gravity and quantum physics becomes possible.

Equation (042) shows the minimum energy required to store one Nat. Given the current temperature of the universe $\mathcal{T} = 1.751\ 10^{-29}\ K$, it is exactly $E_{min} = 2.418\ 10^{-52}\ J$. The updated information from the past is the realized microstate selected from among the multitude of possible microstates. According to Tab. 2, one past second contains $1.501\ 10^{104}$ Nats. That's why a minimum of $E_{1s} = 2.418\ 10^{-52}\ \ 1.501\ 10^{104}\ J = 3.628\ 10^{52}\ J$ of energy is required to store one second's worth of current information.

Chapter 13 demonstrates the mass and energy loss (129) of the universe. In one second, it is $\Delta E = 1.583\ 10^{70}\ 2.293\ 10^{-18}\ J = 3.628\ 10^{52}\ J$.

This is a phenomenal result: The energy lost over time is equal to the energy required to store past information. This explains the constant entropy of the universe (see the end of Chapter 13)!

[23] Andreas Gimsa, The Proof of Mass Decay, International Journal of Scientific Research and Management, Volume 09, Issue 02, February 2021, DOI: 10.18535/ijsrm/ v9i2.aa01, p.54

10. Time as information arises during the Expansion of the Universe

Is there even such a thing as the present? As you read this, what you have just read is already in the past, and what you haven't read yet lies in the future.

My wife had the following opinion on this on September 7, 2025: There must be a present in the form of the tiniest unit of time. Events happen in the present; in the past, they have already happened, and in the future, they are yet to happen. If there were no present, no events could take place at all. I find that convincing.

Time arises from changes in surface area. These occur continuously as mass-bearing objects expand with the expansion of the universe. The sum of all changes in the surface area of the objects leads to an increase in the spherical surface area of the universe. The resulting time possesses an information equivalent. During one oscillation of the universe, the change in surface area occurs twice: once during expansion and once during contraction. Since only expansion is considered here, only half of the surface area change of the spherical universe can be taken into account for the formation of time. It must be noted that the radius increases at the speed of light.

$$I_{uni} = \frac{dA_O}{dt} = 4\pi\frac{dr^2}{dt} = 4\pi\frac{d(c^2t^2)}{dt} \qquad (067)$$

Differentiating (067) yields the following simple result:

$$I_{1\,MZ\,uni} = 8\pi\, t_{uni}c^2 \qquad (068)$$

The total information I_{uni} is divided into a past updated portion $I_{uni\,akt}$ and a future portion $t_{uni\,zuk}$.

$$I_{uni} = I_{akt} + I_{zuk} \qquad (068a)$$

$$\boldsymbol{I_{uni\,akt} = 4\pi t_{uni\,akt}c^2} \qquad (068b) \qquad \boldsymbol{I_{uni\,zuk} = 4\pi t_{uni\,zuk}c^2} \qquad (068c)$$

Equations (068b) and (068c) represent updated and future information that arises as the universe expands. Since the amount of past information increases, according to Equation (068a), the information that will be updated later and lies in the future must become smaller. Future information is updated as the universe expands and becomes correspondingly smaller (see equation (081)).

The current information of a system can only ever be a part of the potential information it contains. It is not the case that the potential information becomes smaller as it is updated from potential information. It remains constant as long as no entropy is transported across the system boundary.

The universe is such a system from which no potential information can escape, since its radius grows exactly at the speed of light.

How is it possible to express the current information depicted in Equation (068b) in Nats? This requires the smallest possible time $t_{uni\,s}$ of the universe. The number of these smallest possible Nat time units is then the current information as an address length. It can then be written using the current age of the universe:

$$I_{Nat\,1\,MZ\,uni} = \frac{4\pi\, t_{uni}\, c^2}{4\pi\, t_{uni\,s}\, c^2} = 8.091\ 10^{60}\ Nats \tag{069}$$

In Chapter 9, the information of the universe required to store all microstates was determined in Nats. This is shown in Equation (064).

Given that the universe is 13.8 billion years old, the number of microstates of the universe's final state is 8.091 10^{60}. This is only a small fraction of the total number of microstates at the end of expansion, as given by Equation (061). However, this information is stored immutably over time: encouraging for all do-gooders and frustrating for all evildoers in this world. Nature does not forget the truth!

A connection between the current information and magnetism in the universe is to be established here. In this regard, it should be noted that in Chapter 16, the electromagnetic cause of gravity is derived.

How high was the **magnetic diffusivity** at the beginning of the universe's expansion? High diffusivity implies rapid propagation of magnetic fields. When potential information is updated in the ether, information must be transported. If the field can be stored—as is common in conductors, for example—the magnetic diffusivity decreases. It is enormous in the universe because the field propagates in a vacuum.

Let us consider the universe as a magnetically conductive sphere.

The magnetic diffusivity of the universe $\eta_{uni\,mg}$ can be calculated using the magnetic field constant μ_0 and the specific conductivity σ_0 or the specific resistance ϱ_0. It should be noted here that the specific electrical conductivity corresponds to the specific magnetic resistance.

$$\eta_{uni\,mg} = \frac{1}{\mu_0\, \sigma_0} = \frac{\varrho_0}{\mu_0} \tag{070}$$

Just like the current information in (068b), the magnetic diffusivity has the unit [m²/s].

What is the resistance of a conducting sphere between its center and its surface? The magnetic current is assumed to flow radially symmetrically from a central minisphere to the surface, as the two poles.

The flux area is $A = 4\,\pi\,r$. The sphere can be decomposed into concentric spherical shells of thickness dr. For the differential resistance dR, the following applies:

$$dR = \varrho_0 \frac{dr}{A(r)} = \varrho_0 \frac{dr}{4\pi\,r^2} \tag{071}$$

The middle small spherical shell with radius r_{min} is supposed to be as large as the radius of the universe at the start of the expansion. Integration over all concentric spherical shells of thickness dr leads to the following result:

$$R = \int_{r_{min}}^{r_{max}} \varrho_0 \frac{dr}{4\pi\,r^2} = \frac{\varrho_0}{4\pi} \int_{r_{min}}^{r_{max}} \frac{dr}{r^2} = -\frac{\varrho_0}{4\pi} \frac{1}{r}\bigg|_{r_{min}}^{r_{max}} \tag{072}$$

With the integration limits, the resistance R of the universe is given by:

$$R = \frac{\varrho_0}{4\pi} \left(\frac{1}{r_{min}} - \frac{1}{r_{max}} \right) \tag{073}$$

For the resistance, the reciprocal of the maximum radius of the universe today can be neglected, since it is very small.

$$R = Z_0 = \frac{\varrho_0}{4\pi} \frac{1}{r_{min}} \qquad\qquad \varrho_0 = 4\pi\, Z_0 r_{min} \tag{074}$$

The resistance R of the universe is the characteristic impedance Z_0 of vacuum. Let us now return to the magnetic diffusivity according to Equation (070):

$$\eta_{uni\,mg} = \frac{\varrho_0}{\mu_0} = \frac{4\pi\, Z_0\, r_{min}}{\mu_0} \tag{075}$$

Using the known relationship $Z_0 = \mu_0\, c$, we can write for (075):

$$\eta_{uni\,mg} = 4\pi\, c\, r_{min} \tag{076}$$

The smallest radius of the universe has already changed at the speed of light. Consequently, the following holds:

$$\eta_{uni\,mg} = 4\pi\, t_{min}\, c^2 \tag{077}$$

What emerges here is the smallest possible amount of information at the beginning of the universe's expansion. It can be interpreted as magnetic diffusivity. Thus, we observe the smallest amount of information in the universe when we measure the magnetic diffusivity of the vacuum.

The radial magnetic current flowing outward from the object masses toward the event horizon of the universe could be the information flow of time. The information may reach the event horizon—as the surface of the universe—in the form of a radial magnetic current for storage.

To obtain the current address length in Nats, one can simply divide Equation (068) by the magnetic diffusivity of the universe:

$$I_{Nat\,uni} = \frac{4\pi\, tc^2}{\eta_{uni\,mg}} \qquad (078)$$

In principle, in addition to the Boltzmann entropy for the current information with its current microstates W_{akt} , the following new formula could apply:

$$I_{akt} = \eta_{uni\,mg}\, ln\, W_{akt} \quad \text{or} \quad W_{akt} = e^{I_{akt}/\eta_{uni\,mg}} \qquad (078a)$$

In the calculation performed so far, it has not been discussed that the radius of the universe expands at the speed of light. In fact, the calculation is suitable for a rigid object in which the magnetic flux flows between its rigid inner sphere and its rigid spherical surface. On the one hand, the calculation of the resistance$R$ considers a rigid object in which the field propagates at the speed of light. On the other hand, the calculation is also intended to apply to a universe whose radius moves at the speed of light. However, these are comparable conditions in which the field moves at the speed of light, just like the radius of the universe. Therefore, in my opinion, the radially symmetric spherical shell model is suitable for calculating the magnetic diffusivity. For the calculation, the dynamic reality was, so to speak, transformed into a static model.

It is possible that the growth of the universe's radius involves the expansion of a magnetic spherical field. This would support the assumption that the formation of time with an increased surface area is a magnetic phenomenon.

Let us take a calculation example for the current information: How many Nats are contained in one hour $(3{,}600\ s)$ of today's universe? According to Table 2, Nos. 1 and 3, one hour $6.677\ 10^{46}$ contains Nats. A test calculation should be performed for comparison. How great is the mass reduction of the universe in one hour?

Equation (044) provides the calculation formula for this. The time difference Δt is 3,600 seconds. Consequently, the mass loss of the universe in one hour is:

$$\Delta m = \frac{3{,}600}{4.362\ 10^{17}}\ 1.761\ 10^{53}\ kg = 1.453\ 10^{39}\ kg \qquad (079)$$

If one hour contains $6.677\ 10^{46}$ nats, one should also be able to calculate the mass loss by multiplying the smallest possible mass loss—calculated for the smallest possible time interval—by the number of nats in an hour:

$$\Delta m_{1h} = 6.677\ 10^{46} \left| \frac{\Delta t_{min}}{t_{uni}}\ m_{uni} \right| \qquad (080)$$

$$\Delta m = 6.677\ 10^{46}\ \frac{5.391\ 10^{-44}}{4.362\ 10^{17}}\ 1.761\ 10^{53}\ kg = 1.453\ 10^{39}\ kg$$

The results of equations (079) and (080) match. This means it is very likely that the calculation is correct.

The future has not yet been updated and is therefore indeterminate. During the expansion phase of the universe, the potential information encoding the address length is updated.

The future is an ever-shrinking portion of the potential information that has not yet been actualized. Consequently, the following applies to future information $I_{Nat\,future}$, with entropy $S_{Nat\,uni}$ and current information $I_{Nat\,uni}$:

$$I_{Nat\,future} = S_{Nat\,uni} - I_{Nat\,uni} \tag{081}$$

As the radius of the universe increases, all periods and time spans must also become larger over time. Consequently, every second also becomes longer. How can one best visualize this?

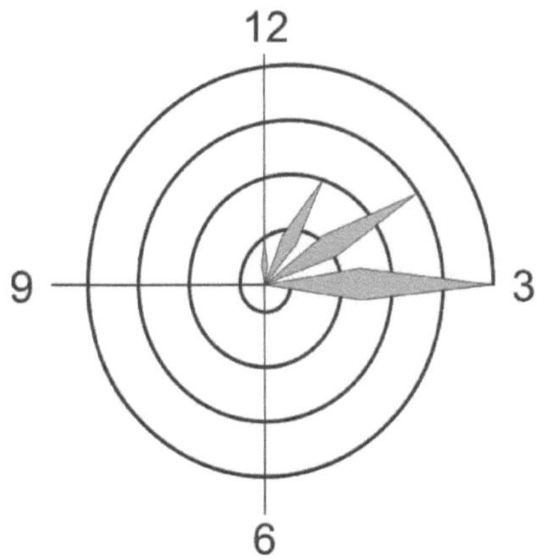

Figure 5: Symbolic clock of the universe with a time spiral

Figure 5 symbolically shows a spiral-shaped clock. The length of a specific time period is expressed by the length of the hand. The hand rotates around the center of the spiral over time and becomes longer with each rotation. In the process, each time period also becomes longer. At first, the still-small hand is at 12; then, after one rotation and 30°, it is at 1; after another rotation and 30°, it is at 2; and after a third rotation and 30°, it is at 3... By the time it reaches the number 3, the length of the hand has grown significantly.

But what does an atomic clock measure? Most atomic clocks use the isotope Cs 133 with a hyperfine structure frequency of $f_{hyp} = 9{,}192{,}631{,}770\text{ Hz}$. The $9{,}192{,}631{,}770 - \text{fache}$ of the period of this frequency corresponds to one second.

A new second begins when the counting of the periods of the previous second is complete. A controlled microwave resonator continuously provides photons with the appropriate energy to maximize the number of energy level transitions in cesium isotopes.

The hyperfine structure frequency is a measure of the energy difference $f = \Delta E / h$ between the two hyperfine ground states of the cesium isotope with total angular momenta $F = 3$ and $F = 4$. In this way, a stable clock is realized.

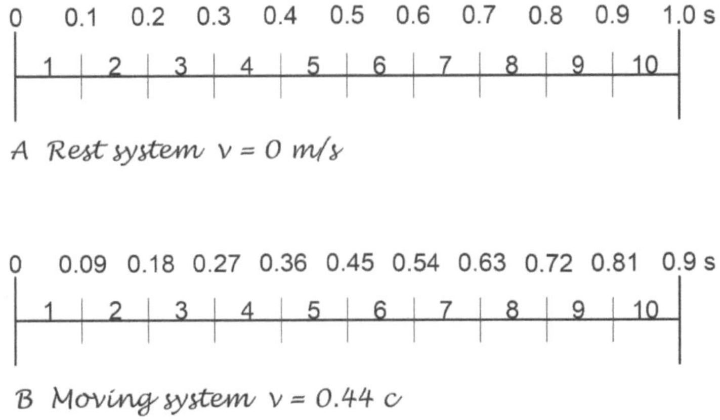

Figure 6: Comparison of clock readings in the rest frame and the moving frame

Long periods of time are made up of short ones. If a long period of time changes due to time dilation in a moving system, all short periods of time must also change proportionally. If one counts the same number of periods (e.g., 10) in a rest frame compared to a moving frame and chooses the rest frame as the reference standard, one finds that the total time span in the moving frame is shorter. This is illustrated in Figure 6. Here, the total time span consisting of 10 periods is 1 second in the rest frame and 0.9 seconds in the moving frame (see also Equation (008)). The moving frame is assumed to be moving at 44% of the speed of light.

Now, in a cesium atomic clock in the rest frame, the second is not composed of 10 periods, but—for high accuracy—of as many as 9,192,631,770 periods. Accordingly, at 44% of the speed of light, the hyperfine structure frequency of the atomic clock would increase to $9{,}192{,}631{,}770/0.9\,\text{Hz} = 10{,}214{,}035{,}300\ \text{Hz}$ to indicate one second.

In principle, one can take the atomic clock that was transferred from the rest frame to the moving frame and return it from the moving frame to the rest frame, where it can be compared with a previously synchronized atomic clock.

After one second in the rest frame, the moving atomic clock has only reached 90% of that second. Since this time span is less than one second, less time has elapsed in the moving frame compared to the rest frame.

This makes it clear that energy and mass increase with velocity. As mass increases, so do the oscillation frequencies within the atomic structure.

One cannot simply swap the positions of two observers, each in one of two inertial frames, as Einstein proposed. This so-called mutual time dilation contains a serious mathematical error.[24] If, for example, both frames were at rest and one later moves uniformly at a constant speed, it must be noted that energy was supplied to the moving frame. It is then also heavier, and only here is the flow of time reduced relative to the rest frame. However, if one assumes that each system can be defined as at rest relative to the other during uniform relative motion, it must be countered that such a consideration is only applicable to massless systems, which do not occur in nature. A pointless endeavor that only causes confusion.

Fig. 4 shows the universe expanding with age; Fig. 5 symbolizes the increasing duration of each time period. Would it be possible to determine the lengths, wavelengths, and periods in the universe that were smaller back then?

As already explained in Chapter 6, the Doppler effect results in a greater redshift as distance increases. This redshift is not only a measure of distance but also of the relative velocity of the galaxies.

In response to what I consider to be a misinterpretation of the accelerating expansion of the universe based on the low brightness of distant supernovae, I would argue the following: At that time, the supernovae were much farther away, and thus much smaller in size and, consequently, had much smaller radiating surfaces. Accordingly, we must see them as dimmer today. If the smaller radiating surface is not taken into account, the measured luminosity suggests a much greater distance at which they are supposedly located.

---

[24] Andreas Gimsa, The Phenomenon of Mass, International Journal of Scientific Research and Management, Volume 08, Issue 06, June 2020, DOI: 10.18535/ijsrm/v8i06.aa01

From this, the false conclusion of an accelerated expansion of the universe is drawn. Isn't it amazing how things all fit together?

The loss of mass in the universe is a phenomenon of information theory! Information is updated in the form of time, and this updating comes at the expense of energy.

Time is the simplest ordering system that assigns a specific sequence to all events in every location. The chronology of events is unchangeable once they have occurred.

Current information that is preserved must be stored. We look into the depths of the universe and, in principle, discover events at any distance that, due to the time of flight of light, lie in the past. Since our observation, conversely, also applies to other observers in all locations, this observation is proof of the conservation of current information.

How can the formation of time be explained electromagnetically on the basis of the complete Maxwell equations, which thus also include elementary magnetic charges? It is my conviction that such a process of time formation must exist.

Chapter 17 describes the possible structure of the proton. Within it, there exist two oppositely charged magnetic elementary charges and the electric elementary charge in the form of a positron. To this end, I introduce relationships with the electric charge Q_{el} and the magnetic charge Q_{mg}. The elementary electric charge within it is the antiparticle positrone^{+}. The two elementary magnetic charges could be a north pole p_{NP} and a south pole p_{SP} , with the south pole being the antiparticle of the north pole. Presumably, the two antiparticles—the positron and the magnetic south pole—which move backward in time, are directly involved in the formation of time. Thus, as they move, time does not pass; rather, it comes into being.

$$t = -\frac{d\,Q_{el}}{d\,V_{mg}} \qquad (081a) \qquad\qquad t = -\frac{d\,Q_{mg}}{d\,V_{el}} \qquad (081b)$$

Equations (081a) and (081b) indicate that a change in electric charge with the magnetic potential V_{mg} and a change in magnetic charge with the electric potential V_{el} can give rise to the formation of time. Proving this is extremely difficult. First, using the two equations, the SI units of the electric and magnetic elementary charges $e\,[As]$ and $p\,[Vs]$, as well as the magnetic and electric potentials $V_{mg}\,[A]$ and $V_{el}\,[V]$, each yield the unit of time "second." The negative sign in the equations indicates a time-forming process.

Due to the small distances between the charges in the proton, the potentials would be very high (see Chapter 17). How does the change in charge occur?

As the universe expands, all partial volumes increase, including every charge space and every proton. Since charge conservation applies, the elementary charges are distributed over larger volumes throughout the universe. Over time, this results in a lower charge density, which is equivalent to a reduction in charge within a fixed volume.

Chapter 13 demonstrates that as the universe expands, all objects, including protons, increase in size. Chapter 17 explains the structure of the proton. This makes it apparent that time could, in principle, arise within protons. If the elementary charges are present in the protons as explained in Chapter 17, then the electric and magnetic potentials also exist. Time formation according to Equations (081a) and (081b) is thus possible.

Equation (060) shows that as the age of the universe increases, mass decreases. This also applies to all protons; they become lighter. As age increases, time passes. This raises the question: in a time-forming process, shouldn't mass conversely increase over time at the points where time is formed? And if time formation occurs within protons, shouldn't they then become heavier?

The structure of the proton, as explained in Chapter 17, includes the antiparticles of the electric and magnetic elementary charges. It is only through these that the formation of time becomes possible at all. The evolution of an antimatter particle over time is equivalent to the evolution of the matter particle in the reverse direction of time. Consequently, the antimatter particle becomes heavier in the reverse direction of time, that is, toward the past. This means that in the normal direction of time, from the past to the future, there is no increase in mass, but rather a reduction in mass.

Finally, an answer should be given as to what time actually is. Equation (068) has shown that it is a type of information and possesses an information equivalent. What kind of information is this? The age of the universe is the current address information of the universe; it specifies the address length required for the current state space to address a microstate. The current address length—and with it the current address space—grows exorbitantly with every second up to the maximum potential number of address-Nats, which results from entropy.

Consequently, every new time interval represents an increase in the address length of the universe and thus an increase in address-Nats. Consequently, there are always more possible events available in the universe.

11. The Correct Relativistic Conservation of Momentum

Equation (082) shows the well-known formula for relativistic kinetic energy:

$$E_{1kin} = \left(\frac{1}{\sqrt{1-\frac{v_1{}^2}{c^2}}} - 1\right) m_{10}c^2 \qquad E_{2kin} = \left(\frac{1}{\sqrt{1-\frac{v_2{}^2}{c^2}}} - 1\right) m_{20}c^2 \qquad (082)$$

Equation (083) shows the well-known energy-momentum relation, which is said to be unsuitable for relativistic calculations.

$$2m_{10}E_{1kin} = p_1{}^2 \qquad 2m_{20}E_{2kin} = p_2{}^2 \qquad (083)$$

Upon closer examination, however, it turns out that Equation (083) is correct and covers the relativistic cases as well. Indeed, Equation (082) can be substituted into (083), yielding the correct relativistic conservation of momentum, which thus follows directly from the relativistic kinetic energy. For a two-mass system, the following applies to the magnitudes of the momenta:

$$|p_1| = |p_2| \qquad m_{10}\, c \sqrt{\frac{2}{\sqrt{1-\frac{v_1{}^2}{c^2}}} - 2} = m_{20}\, c \sqrt{\frac{2}{\sqrt{1-\frac{v_2{}^2}{c^2}}} - 2} \qquad (084)$$

Although the well-known relativistic momentum conservation in the form $\frac{m_{10}}{\sqrt{1-\frac{v_1{}^2}{c^2}}} v_1 = \frac{m_{20}}{\sqrt{1-\frac{v_2{}^2}{c^2}}} v_2$ leads to results similar to those of equation (084) in many applications, it is not physically or mathematically identical to it and is therefore, in my opinion, incorrect.

Both Equation (082) and Equation (083) are well-known, have been extensively tested in practice, and are both correct. Equation (084) is simply the logically inevitable mathematical consequence of the two preceding equations. Equation (084), just like the energy equation (082), also shows a velocity-dependent increase in momentum, in which the object cannot reach the speed of light.

While in classical and the well-known relativistic conservation of momentum, the total momentum can be defined as the sum of two individual momenta equaling zero due to the opposite signs of the velocities, this is not possible according to (084). Each individual momentum retains a positive sign.

Since the Lorentz transformation applies only to massless objects and is therefore unnatural, the derivation of relativistic momentum conservation is also incorrect. Not even its use as an approximate equation is recommended in light of equation (084).

12. The relationship between gravitational and kinetic energy [25]

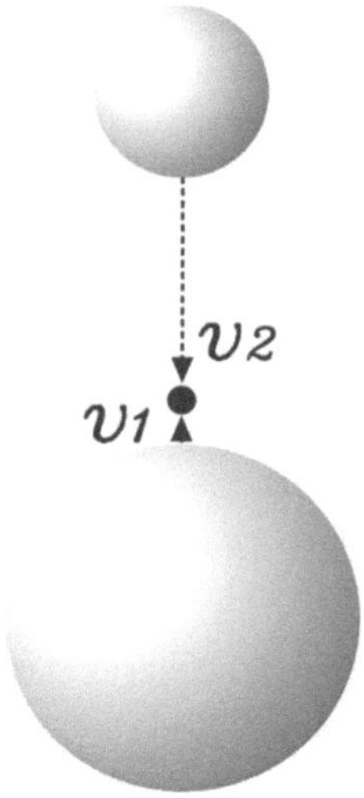

Figure 7: Translation

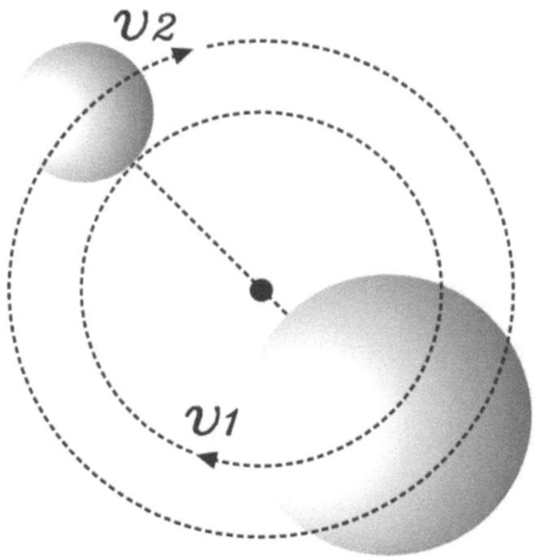

Figure 8: Rotation

The gravitational energy, as potential energy in the rest state of two masses m_{10} and m_{20} before they fall freely toward each other, corresponds exactly to the total kinetic energy at their collision. To calculate the gravitational energy for this, the gravitational force must be integrated over the path up to the collision of the masses. In the rest state, the masses have a center-to-center distance r , which reduces to the distance $r_1 + r_2$ upon collision of the surfaces:

$$E_{grav} = \int_{r_1+r_2}^{r} F_{grav}\, ds = \int_{r_1+r_2}^{r} -G\,\frac{m_{10}\, m_{20}}{s^2}\, ds = G\,\frac{m_{10}\, m_{20}}{s}\bigg|_{r_1+r_2}^{r} \qquad (085)$$

Here, r is the center-to-center distance between the two masses, and r_1 and r_2 are the radii assuming spherical objects.

$$E_{grav} = -G\, m_{10}\, m_{20}\left(\frac{1}{r_1+r_2} - \frac{1}{r}\right) = -G\,\frac{m_{10}\, m_{20}}{r_b} \qquad (086)$$

The term involving the radii of the objects r_1 and r_2 as well as the center-to-center distance r can be defined as r_b for easier handling:

$$r_b = \frac{r(r-h)}{h} = \frac{1}{\frac{1}{r_1+r_2} - \frac{1}{r}} \qquad (087)$$

[25] Andreas Gimsa, Two-Mass Systems Under the Influence of Gravity, Books on Demand, Norderstedt, 2017, ISBN 978-3-00-057218-0

Here, $h = r - (r_1 + r_2)$ is the drop height, i.e., the surface distance between the two masses before the fall.

The quantity r_b, which includes the center-to-center distance and the object radii, seems somewhat cumbersome. To give the mathematically inclined reader a little treat, we briefly examine the special case for which $r_b = h$ holds:

$$r_b = h = \frac{r(r-h)}{h} \qquad r^2 - rh - h^2 = 0 \tag{088}$$

$$r_{1/2} = \frac{h}{2} \pm \sqrt{\frac{h^2}{4} + \frac{4}{4}h^2} = \frac{h}{2} \pm \sqrt{\frac{5}{4}h^2} \qquad r_{1/2} = h\frac{1 \pm \sqrt{5}}{2} \tag{089}$$

The positive of the two solutions shows that if the ratio of the distance from the center to the surface r/h corresponds to the golden ratio, the quantity r_b becomes the height of fall h . Such a two-mass system is entirely conceivable.

As is well known, the kinetic energies can be interpreted as increases in mass Δm relative to their rest masses (see (082)). Using Equation (087), we can write:

$$|E_{grav}| = E_{1kin} + E_{2kin} = G\frac{m_{10}\, m_{20}}{r_b} = \Delta m_1 c^2 + \Delta m_2 c^2 \tag{090}$$

For the rotation of two masses in an orbit around their common center of mass, a similar relationship applies:

$$|E_{grav}| = 2E_{1kin} + 2E_{2kin} = G\frac{m_{10}\, m_{20}}{r} = 2\Delta m_1 c^2 + 2\Delta m_2 c^2 \tag{091}$$

Why does the kinetic energy in rotational equilibrium according to Equation (091) appear with a factor of 2 compared to translational motion according to Equation (090)? The answer is simple, yet it remained a mystery to me for a long time. The translation under consideration is only a half-cycle. We are considering the free fall of two masses that fall toward each other from a certain distance and their respective rest positions. Strictly speaking, the collision is only the turning point during a perfectly elastic impact. A frictionless oscillation would be complete once the masses have returned to their rest positions.

What can be written for each of the two kinetic energies so that their sum equals the gravitational energy $|E_{grav}| = G\frac{m_{10}\, m_{20}}{r_b}$?

$$E_{1kin} = \Delta m_1 c^2 = G\frac{m_{10}\, m_{20}}{r_b}\frac{m_{20}}{m_{10}+m_{20}} \tag{092}$$

$$E_{2kin} = \Delta m_2 c^2 = G\frac{m_{10}\, m_{20}}{r_b}\frac{m_{10}}{m_{10}+m_{20}} \tag{093}$$

To prove this, let's test it using the gravitational energy E_{grav} :

$$|E_{grav}| = E_{1kin} + E_{2kin} = G\frac{m_{10}\,m_{20}}{r_b}\left(\frac{m_{20}}{m_{10}+m_{20}} + \frac{m_{10}}{m_{10}+m_{20}}\right) \qquad (094)$$

Equation (094) can be easily simplified, leaving the gravitational energy:

$$|E_{grav}| = G\frac{m_{10}\,m_{20}}{r_b}\left(\frac{m_{10}+m_{20}}{m_{10}+m_{20}}\right) = G\frac{m_{10}\,m_{20}}{r_b} \qquad (095)$$

Why is the unusual notation for kinetic energy introduced in (092) and (093)? It is needed later to calculate the time dilations and velocities during the collision of the two objects after they fall from their respective rest positions toward each other.

The conservation of entropy, which holds true for the universe and can be declared a universal law, takes the form given by Equation (060):

$S = \frac{k}{\hbar} m_{uni}\, t_{uni}\, c^2$. Entropy remains constant as the universe expands.

It will be shown that the principle of entropy conservation also applies to gravity!

What does Equation (060) mean? It means nothing less than that as the universe ages, every mass in the universe must become smaller. That is, all galaxies, stars, and celestial bodies are constantly becoming lighter! What exactly happens with gravity? Exactly the same thing, only in reverse: As masses attract each other, their velocities increase. This slows down the flow of time, so that every time interval on every mass becomes shorter as its speed increases, and at the same time, this mass grows due to its speed. Does this contradict the Special Theory of Relativity? Yes, of course, because there, time dilation is considered with massless inertial systems that are in relative motion to one another. Alternatively, each observer can claim the same point of view on one of two massless objects in each system and assert that the other's time always passes more slowly. This is nonsense. For it fails to consider that an object moving at high speed possesses greater energy than in its rest state, and therefore the time base changes at that speed. The relativistic mass increase is correct in and of itself. However, in the Special Theory of Relativity, it is not placed in the necessary context with time dilation. For, according to Equation (060), the relationship between the object at rest and the object in motion holds for every mass:

$$m_{10}t_{10} = m_{1b}t_{1b} \qquad (096) \qquad\qquad m_{20}t_{20} = m_{2b}t_{2b} \qquad (097)$$

In equations (096) and (097), the masses m_0 denote rest masses, the times t_0 denote time intervals in the rest state, the masses m_b denote masses in motion, and the times t_b denote time intervals in the moving state.

The term $m_{10}t_{1b}$ is now subtracted from both sides of equation (096), and the term $m_{20}t_{2b}$ is now subtracted from both sides of equation (097):

$$m_{10}t_{10} - m_{10}t_{1b} = m_{1b}t_{1b} - m_{10}t_{1b} \tag{098}$$

$$m_{20}t_{20} - m_{20}t_{2b} = m_{2b}t_{2b} - m_{20}t_{2b} \tag{099}$$

Using equations (098) and (099), we can now introduce the following notation:

$$\Delta t_1 m_{10} = -\Delta m_1 t_{1b} \quad (100) \qquad \Delta t_2 m_{20} = -\Delta m_2 t_{2b} \quad (101)$$

$$\frac{\Delta t_1}{t_{1b}} = -\frac{\Delta m_1}{m_{10}} \quad (102) \qquad \frac{\Delta t_2}{t_{2b}} = -\frac{\Delta m_2}{m_{20}} \quad (103)$$

Using equations (092) and (093), we can now easily write equations (104) and (105):

$$\frac{\Delta m_1}{m_{10}} = \frac{G\, {m_{20}}^2}{c^2 r_b (m_{10}+m_{20})} \quad (104) \qquad \frac{\Delta m_2}{m_{20}} = \frac{G\, {m_{10}}^2}{c^2 r_b (m_{10}+m_{20})} \quad (105)$$

For the rotation of a two-mass system, the following applies, derived from equation (091) in a similar manner:

$$\frac{\Delta m_1}{m_{10}} = \frac{G\, {m_{20}}^2}{2\, c^2 r (m_{10}+m_{20})} \quad (106) \qquad \frac{\Delta m_2}{m_{20}} = \frac{G\, {m_{10}}^2}{2\, c^2 r (m_{10}+m_{20})} \quad (107)$$

And from equations (102) and (106) as well as (103) and (107), the correct specific time dilation now follows directly; this refers here to the time interval t_b, i.e., the time interval that elapses on the moving object. From this, using equations (092) and (093), the principle also follows that the specific time dilation $\Delta t/t_b$ on a mass can also be regarded as the ratio of the kinetic to the rest energy of that mass. I consider this remarkable, and it has not yet been published in this form:

$$\frac{\Delta t_1}{t_{1b}} = -\frac{G\, {m_{20}}^2}{c^2 r_b (m_{10}+m_{20})} = -\frac{E_{1kin}}{E_{10}} \tag{108}$$

$$\frac{\Delta t_2}{t_{2b}} = -\frac{G\, {m_{10}}^2}{c^2 r_b (m_{10}+m_{20})} = -\frac{E_{2kin}}{E_{20}} \tag{109}$$

For the rotation of a two-mass system, derived from equation (091):

$$\frac{\Delta t_1}{t_{1b}} = -\frac{G\, {m_{20}}^2}{2\, c^2 r\, (m_{10}+m_{20})} = -\frac{\Delta m_1}{m_{10}} \tag{110}$$

$$\frac{\Delta t_2}{t_{2b}} = -\frac{G\, {m_{10}}^2}{2\, c^2 r\, (m_{10}+m_{20})} = -\frac{\Delta m_2}{m_{20}} \tag{111}$$

If we now divide equation (104) by (105), we arrive at a surprising result for the two-mass system.

The ratio of the mass increases of both masses, which represent the kinetic energies, corresponds to the inverse mass ratio of both rest energies:

$$\frac{\Delta m_1}{\Delta m_2} = \frac{E_{1kin}}{E_{2kin}} = \frac{m_{20}}{m_{10}} = \frac{E_{20}}{E_{10}} \qquad (112)$$

From the relationship between gravitational energy and kinetic energy according to equations (094) and (095), the equality between velocity-induced and gravitational time dilation now holds for equations (110) and (111).

In other words, the following principle applies in the two-mass system: The time dilation that an object experiences on the surface of a second object due to the latter's gravitational field is equal to the velocity-induced time dilation it would experience if it were to reach the surface of the second object in free fall from rest. This principle holds reciprocally.

A comparison of the time dilation according to Equation (111) with that from the theory of relativity leads to better agreement with the well-known experiments Pound/Rebka 1960, Hafele/Keating 1971, Maryland 1976, Briatore/Leschiutta 1976, Chou/Hume/Rosenband/Wineland 2010. [26]

In a two-mass system, there are always two time dilations during free fall: $\Delta t_1/t_{1b}$ and $\Delta t_2/t_{2b}$ —one on the first mass due to its velocity and one on the second mass due to its velocity. The velocities of both masses, and thus also the time dilations, follow directly from the gravitational interaction between the two objects.

<u>We now need to determine the relativistic velocities at the collision of the two objects, which collide in free fall from rest at a distance of r :</u>

Using the time dilation from Equation (008), the mass-time relations from Equations (096) and (097), and also the derivation from the ether momentum in Equation (029), the following applies to the moving masses:

$$m_{1b} = \frac{m_{10}}{\sqrt{1-\frac{v_1^2}{c^2}}} \qquad (113) \qquad m_{2b} = \frac{m_{20}}{\sqrt{1-\frac{v_2^2}{c^2}}} \qquad (114)$$

Next, the rest mass is subtracted from both sides of the equations:

$$m_{1b} - m_{10} = \frac{m_{10}}{\sqrt{1-\frac{v_1^2}{c^2}}} - m_{10} \qquad (115) \qquad m_{2b} - m_{20} = \frac{m_{20}}{\sqrt{1-\frac{v_2^2}{c^2}}} - m_{20} \qquad (116)$$

[26] Andreas Gimsa, The Gravitational Time Dilation of a Two-Mass System, Part 8, Kuss GmbH, Potsdam, 2017, ISBN 978-3-00-056508-3

It then follows immediately that:

$$\Delta m_1 = m_{10}\left(\frac{1}{\sqrt{1-\frac{v_1^2}{c^2}}} - 1\right) \quad (117) \qquad \Delta m_2 = m_{20}\left(\frac{1}{\sqrt{1-\frac{v_2^2}{c^2}}} - 1\right) \quad (118)$$

Further rearrangement of equations (117) and (118) and comparison with (104) and (105) leads to the following equations:

$$\frac{G\, m_{20}^2}{c^2 r_b (m_{10}+m_{20})} + 1 = \frac{1}{\sqrt{1-\frac{v_1^2}{c^2}}} \quad (119) \qquad \frac{G\, m_{10}^2}{c^2 r_b (m_{10}+m_{20})} + 1 = \frac{1}{\sqrt{1-\frac{v_2^2}{c^2}}} \quad (120)$$

From this, the relativistic velocities v_1 and v_2 can be easily derived:

$$v_1 = c\sqrt{1 - \frac{1}{\left[\frac{G\, m_{20}^2}{c^2 r_b (m_{10}+m_{20})}+1\right]^2}} \quad (121)$$

$$v_2 = c\sqrt{1 - \frac{1}{\left[\frac{G\, m_{10}^2}{c^2 r_b (m_{10}+m_{20})}+1\right]^2}} \quad (122)$$

These velocities account for both momentum conservation and energy conservation for both objects. These calculation rules have already been published by me.[27] The formula $v = \sqrt{2\, g\, h}$ commonly found in physics textbooks—which is perfectly suitable for falling masses that are small compared to the Earth's mass and do not fall from very great heights—suggests that the mass of the falling body plays no role in its velocity. This is incorrect.

For the rotation of a two-mass system, the following applies to the two velocities of two masses rotating about their common center of mass:

$$v_1 = c\sqrt{1 - \frac{1}{\left[\frac{G\, m_{20}^2}{2\, c^2 r\, (m_{10}+m_{20})}+1\right]^2}} \quad (123)$$

$$v_2 = c\sqrt{1 - \frac{1}{\left[\frac{G\, m_{10}^2}{2\, c^2 r\, (m_{10}+m_{20})}+1\right]^2}} \quad (124)$$

Here, the factor 2 discussed earlier reappears in the equations for rotation versus translation; this is due to the fact that, in rotation, energy conservation is considered for a complete oscillation, whereas in translation, it is considered only for half an oscillation up to the impact.

[27] Andreas Gimsa, Two-Mass Systems Under the Influence of Gravity, Books on Demand, Norderstedt, 2017, ISBN 978-3-00-057218-0, p. 13

The reader is expressly encouraged to verify these relativistic relationships, which take into account the constancy of entropy in gravitation according to equations (096) and (097). For example, one can simply compare calculation results for translation involving stones of different masses falling to Earth with experimental data. To verify the rotational velocities around the common center of mass of a two-mass system, the Earth-Moon system or any binary star system is suitable. This can be compared with astronomical observations and known data.

13. The Change in "Constants of Nature"

I am aware that the idea of various physical quantities changing with the age of the universe takes some getting used to. A typical skeptical comment on this is as follows: "The author is asking us to believe, contrary to current knowledge, that all masses, periods, the gravitational constant, etc., are changing." No, I do not. However, I hope that interested, serious scientists will engage with the proven changes and examine them. This will give them the opportunity to expand their understanding of physics and to use the knowledge I have published for their research. For these changes are indispensable and of fundamental importance to our understanding of the world. They will now be explained in more detail.

To lay some groundwork, I will first list important constants:

$\hbar$	*Planck constant*
k	*Boltzmann constant*
σ	*Stefan-Boltzmann constant*
c	*Speed of light*
e	*Elementary electric charge*
p	*Magnetic elementary charge*
ε_0	*Electric field constant*
μ_0	*Magnetic field constant*
α	*Fine-structure constant*
κ	*Isentropic exponent of the universe* ($\kappa = 4/3$)

All of these physical quantities are as unchanging as the number π.

As the universe ages, typical velocities v_x —such as the rotational velocities of stars in galaxies—and typical resistances R_x —such as the wave resistance of a vacuum or the Klitzing resistance—remain constant.

As already explained in Chapter 7, an isentropic expansion of the universe is taking place. All masses m_x change with the mass m_{uni} and the age of the universe t_{uni}. In this process, the product of mass and age remains constant, as shown by Equation (125):

$$m_{uni}\, t_{uni} = m_{uni\,neu}\, t_{uni\,neu} \qquad m_x\, t_{uni} = m_{x\,neu}\, t_{uni\,neu} \tag{125}$$

$$m_{x\,neu} = \frac{m_x\, t_{uni}}{t_{uni\,neu}} \tag{126}$$

Any future time interval Δt can be defined relative to the current age t_{uni} of the universe:

$$\Delta t = t_{uni\,neu} - t_{uni} \tag{127}$$

Equation (126) can be rewritten using (127), yielding the new reduced mass that exists after the time interval Δt:

$$m_{x\,new} = \frac{m_x\,t_{uni}}{t_{uni}+\Delta t} \qquad m_{x\,new} = \frac{m_x}{1+\frac{\Delta t}{t_{uni}}} \tag{128}$$

The actual change in mass Δm_x over the time interval Δt can now also be easily calculated:

$$\Delta m_x = m_x - m_{x\,new} = m_x - \frac{m_x}{1+\frac{\Delta t}{t_{uni}}} = m_x\left(1 - \frac{1}{1+\frac{\Delta t}{t_{uni}}}\right) \tag{129}$$

Changes in energy (E_x), frequency (f_x), and temperature ($\mathcal{T}_x$) must be proportional to the changes in mass:

$$E_{x\,new} = \frac{E_x}{1+\frac{\Delta t}{t_{uni}}} \quad (130) \qquad f_{x\,new} = \frac{f_x}{1+\frac{\Delta t}{t_{uni}}} \quad (131) \qquad \mathcal{T}_{x\,new} = \frac{\mathcal{T}_x}{1+\frac{\Delta t}{t_{uni}}} \quad (132)$$

Since all masses decrease, all energies, frequencies, and temperatures also decrease. Since these quantities all possess energy content, they must change accordingly to preserve the cosmic structure.

However, all distancess_x also change as follows with the age and expansion of the universe:

$$\frac{r_{uni}}{t_{uni}} = \frac{r_{uni\,new}}{t_{uni\,new}} = c \qquad \frac{s_x}{t_{uni}} = \frac{s_{x\,new}}{t_{uni\,new}} \qquad s_{x\,neu} = s_x\frac{t_{uni\,new}}{t_{uni}} \tag{133}$$

$$s_{x\,new} = s_x\frac{t_{uni}+\Delta t}{t_{uni}} \qquad s_{x\,neu} = s_x\left(1 + \frac{\Delta t}{t_{uni}}\right) \tag{134}$$

The expansion of the universe thus goes far beyond the increase in the distances between galaxies (see Chapter 6).

Among the distances that change are, of course, the Bohr radius and the proton radius.

As a verifiable cosmological example, the annual increase in the distance between the Earth and the Moon can be precisely calculated using the variable quantities of orbital radius, mass, and gravitational constant (see Chapter 14).

Here is an impressive formula for calculating the age of the universe, derived from the relationships in Chapters 20 and 21:

$$t_{uni} = \sqrt{\frac{\hbar^3\,r_p}{c\,G^2\,m_e{}^2 m_p{}^3}} \tag{135}$$

Strictly speaking, it follows from the lever law of the universe according to (292) together with (293) and (316). So if you want to know the exact age of the universe, use my formula. What is the exact age that comes out?

The calculation using an average of 31,556,952 s/a yields 13.8 billion years. Can this calculation be trusted? How likely is it that a purely random calculation involving 7 physical quantities—linked via multiplication, division, and exponentiation—would result in the unit of seconds?

Furthermore, how likely is it that such a formula would actually yield the age of the universe? Perhaps we should take the formula seriously.

One thing, however, becomes immediately clear: Since equation (135) is correct, at least one of the 7 physical quantities must be variable, because the age of the universe is guaranteed to be variable. Strictly speaking, 5 of the 7 quantities are variable.

For the change in all periods with the age of the universe t_{uni} , the following applies:

$$T_{x\,new} = T_x\left(1 + \frac{\Delta t}{t_{uni}}\right) \tag{136}$$

Consequently, all areas A_x vary with the square of the distances s_x :

$$A_{x\,new} = A_x\left(1 + \frac{\Delta t}{t_{uni}}\right)^2 \tag{137}$$

All volumes V_x therefore vary with the cube of the distances s_x :

$$V_{x\,new} = V_x\left(1 + \frac{\Delta t}{t_{uni}}\right)^3 \tag{138}$$

All forces F_x and powers P_x must vary as follows due to the changes in energy, distance, and time:

$$F_{x\,new} = \frac{F_x}{\left(1+\frac{\Delta t}{t_{uni}}\right)^2} \tag{139}$$

$$P_{x\,new} = \frac{P_x}{\left(1+\frac{\Delta t}{t_{uni}}\right)^2} \tag{140}$$

Therefore, the gravitational force must also decrease quadratically with the age of the universe. No **dark energy** is required for the expansion; the gravitational force simply decreases quadratically with the age of the universe. This reduces the cohesion of the masses. The gravitational force decreases because the masses become smaller and the distances greater.

What happens to Newton's gravitational constant G? It also changes with the age of the universe. The ratios of the fundamental forces must remain constant. This also applies to the ratio of the gravitational force to the electric force:

$$\frac{F_{grav}}{F_{el}} = \frac{\frac{G\,m_e m_p}{r_{bohr}{}^2}}{\frac{e^2}{4\pi\,\varepsilon_0 r_{bohr}{}^2}} = \frac{G\,m_e m_p}{\frac{e^2}{4\pi\,\varepsilon_0}} = const. \tag{141}$$

The elementary charge e and the electric field constant ε_0 are true natural constants. Since both the electron and proton masses decrease with the age of the universe, the gravitational constant G must increase quadratically:

$$G\, m_e m_p = G_{new}\, m_{e\,new}\, m_{p\,new} = G_{new} \frac{m_e}{1+\frac{\Delta t}{t_{uni}}} \frac{m_p}{1+\frac{\Delta t}{t_{uni}}} = G_{new} \frac{m_e\, m_p}{\left(1+\frac{\Delta t}{t_{uni}}\right)^2} \quad (142)$$

$$\frac{G_{new}}{G} = \left(1 + \frac{\Delta t}{t_{uni}}\right)^2 \quad (143)$$

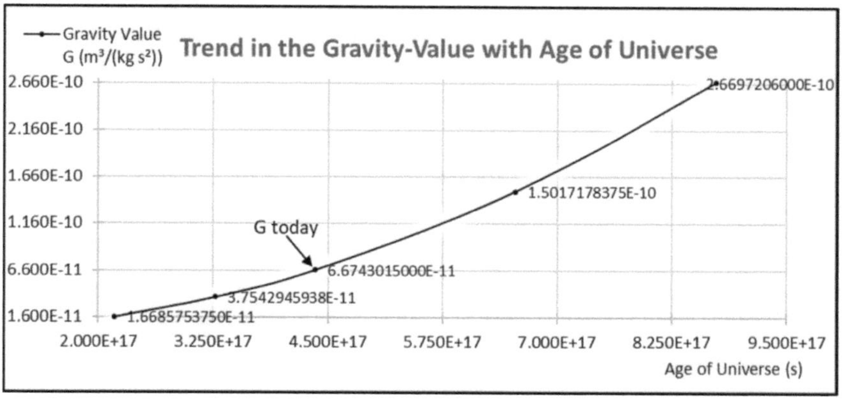

Figure 3: Gravitational constant from half to twice the age of the universe

Figure 3 shows the quadratic evolution of the gravitational constant over time according to Equation (143). The arrow marks the current age of the universe. [28]

Various electrical and magnetic quantities naturally also change with the expansion of the universe. These include the electric and magnetic field strengths, which must change as forces:

$$E_{el\,new} = \frac{E_{el\,x}}{\left(1+\frac{\Delta t}{t_{uni}}\right)^2} \quad (144) \qquad H_{mg\,new} = \frac{H_{mg\,x}}{\left(1+\frac{\Delta t}{t_{uni}}\right)^2} \quad (145)$$

The electric and magnetic flux densities also change like forces:

$$D_{el\,new} = \frac{D_{el\,x}}{\left(1+\frac{\Delta t}{t_{uni}}\right)^2} \quad (146) \qquad B_{mg\,new} = \frac{B_{mg\,x}}{\left(1+\frac{\Delta t}{t_{uni}}\right)^2} \quad (147)$$

All voltages and currents must change in the same way as energies:

$$U_{new} = \frac{U_x}{1+\frac{\Delta t}{t_{uni}}} \quad (148) \qquad I_{new} = \frac{I_x}{1+\frac{\Delta t}{t_{uni}}} \quad (149)$$

[28] Andreas Gimsa, The Beauty of Nature, 2nd edition, Gieselmann Druck und Medienhaus, Potsdam, 2014, ISBN 978-3-923830-97-8, p. 75

For capacitances and inductances, the changes with the age of the universe are inversely proportional to voltages and currents:

$$C_{new} = C_x\left(1 + \frac{\Delta t}{t_{uni}}\right) \qquad (150) \qquad\qquad L_{new} = L_x\left(1 + \frac{\Delta t}{t_{uni}}\right) \qquad (151)$$

While momenta p_x decrease with mass, angular momenta L_x remain constant.

This overview does not claim to be exhaustive. There are many derived physical quantities for which the reader can now consider for themselves whether they must change with the age of the universe.

Almost all **Planck units** are described using the gravitational constant G. The consequence of this is that these Planck units vary with the age of the universe! It's a bit of a shame, as they seemed so stable. But nature doesn't debate this with us; it is an irrefutable fact that constitutes an important prerequisite for its harmony.

I disclosed these findings over 10 years ago. It is truly astonishing that they have not yet found their way into physics. Aside from my limited popularity, it is certainly also because new findings are often inconvenient, as they call existing and established knowledge into question.

All the changes described here assume that the entropy of the universe remains constant. How can entropy remain constant if dissipative structures should lead to an increase in entropy? Where is the energy sink?

It has been proven that systems that are not in thermodynamic equilibrium can develop dissipative structures if there is a continuous flow of energy and matter within them. [29] What might this flow of energy in the universe consist of? The energy and mass of the universe are decreasing, even though it is a closed system. Energy is required to store information that arises over time. Past information is stored in the universe and remains unchangeable. Mass passes into the past without friction losses.

Another flow of energy is initiated by gravity. Potential energy between separated masses is converted into kinetic energy. After two masses collide, the kinetic energy and the negative gravitational energy are greater than before the objects' gravitational motion (see Chapter 12). Until the collision, entropy remains constant even under the influence of gravity.

[29] Gregoire Nicolis, Ilya Prigogine, Self-Organization in Nonequilibrium Systems: From Dissipative Structures to Order through Fluctuations, New York: John Wiley & Sons., 1977, ISBN 978-0-471-59948-6

14. The Variable Distance between Earth and the Moon [30]

The expansion of the universe is more significant than previously assumed. Due to the constancy of entropy, all masses must fall and all distances must increase. This is examined using the example of the evolution of the Earth-Moon distance. The influence of tidal friction on the increase in distance is investigated and quantified. It was found that it exerts a lesser influence than is currently assumed. When the influence of expansion and tidal friction are taken into account, the calculations performed here show very good agreement with the measured distance. The universe is a closed system. No transport of energy or matter can occur across the system boundaries. The expansion of space occurs isentropically, i.e., at constant entropy. Formula (060) applies to the entropy of the universe:

$$S_{uni} = \frac{\mathrm{k}}{\hbar} m_{uni} t_{uni} c^2 = \frac{\mathrm{k}}{\hbar} N\hbar = Nk \tag{152}$$

N is the address length of $6.5\ 10^{121}$ (see Chapter 9). This can be interpreted as potential information of the macrosystem "Universe," since it also determines the number of microstates. The mass m_{uni} of the Universe must decrease with its age t_{uni}. Along with it, the masses of all celestial bodies of any mass m_x decrease. It therefore holds that: $m_x t_{uni} = const$. For the evolution of the masses of the Earth and the Moon m_1 and m_2 over time, it follows that:

$$m_{1b} = \frac{m_{1a}}{1+\frac{\Delta t}{t_{uni\,a}}} \quad (153) \qquad m_{2b} = \frac{m_{2a}}{1+\frac{\Delta t}{t_{uni\,a}}} \quad (154)$$

Boltzmann constant	k	1.38064900E-23	J/K
Planck's constant	$\hbar$	1.05457182E-34	Js
Speed of light	c	2.99792458E+08	m/s
Gravitational constant	G	6.67430000E-11	$m^3/(kg\ s^2)$
Age of the Universe	t_{uni}	4.36000000E+17	s
Mass of the universe	m_{uni}	1.76000000E+53	kg
Earth's mass	m_1	5.97700000E+24	kg
Moon mass	m_2	7.35200000E+22	kg
Earth radius	r_1	6.37100000E+06	m
Earth-Moon center distance	r_a	3.84400000E+08	m
Length of day	T_1	8.64000000E+04	s

Table 3: Input values for the calculation

[30] Andreas Gimsa, Development of the Earth-Moon Distance, International Journal of Scientific Research and Management, Volume 08, Issue 03, March 2020, DOI: 10.18535/ijsrm/v8i03.aa01

The index a indicates the current calculated value, and the index b indicates the calculated value one year from now $(\Delta t = 1a)$.

The following three changes (1 through 3) in physical quantities are significant for calculating the annually increasing Earth-Moon distance:

1. <u>The time base of oscillating systems increases with the age of the universe</u>

As the age of the universe t_{uni} increases, the time base of all periodic processes must change. Since all masses are decreasing, the time base of periodic processes must increase. For example, each photon also becomes less energetic; its mass equivalent and its frequency decrease: $mc^2 = hf = h/T$. In general, as masses decrease, the periods increase:

$$\frac{\mathrm{T}_a}{t_{uni\,a}} = \frac{\mathrm{T}_b}{t_{uni\,b}} = \frac{\mathrm{T}_b}{t_{uni\,a}+\Delta t} \qquad T_b = T_a\left(1 + \frac{\Delta t}{t_{uni\,a}}\right) \tag{155}$$

When the time of flight of light is measured with an atomic clock for a distance measurement, by today's standards, more seconds elapse each year for the light's path, resulting in a greater distance. This distance is then measured as it has actually increased due to the expansion of the universe.

2. <u>The increase in all distances with the age of the universe</u>

For example, for the Bohr radius of the hydrogen atom in its lowest energy state, the relationship $\mathrm{r}_{bohr} = \hbar/(\alpha\ c\ m_e)$ holds. That is, as the mass of the electron decreases, the hydrogen atom must become larger.

This increase in size applies to all atomic distances and separations in connection with equations (153), (154), and (155). They increase with the reduction in mass:

$$r_b = r_a\left(1 + \frac{\Delta t}{t_{uni\,a}}\right) \tag{156}$$

Using (156), one can immediately calculate the extent to which the distance between the Moon and the Earth increases, namely by $\Delta r = 2.8\ cm/a$.

In the two-mass system, the center-to-center distance between the two celestial bodies can be calculated using the common period, both masses, and the gravitational constant, which, as mentioned, is not constant.

This formula is used to verify the result from equation (156):

$$r = \sqrt[3]{\frac{T^2 G(m_1+m_2)}{4\pi^2}} \tag{157}$$

This formula with the common period T in the two-mass system is well known.

3. The evolution of the gravitational constant with the age of the universe

If all masses and periods change with the age of the universe, the question arises as to whether the gravitational constant G also changes over time.

Paul Dirac had already postulated a variable gravitational constant. In his view, however, it was supposed to decrease over time. To preserve the atomic structure, the ratio of the electric force F_e to the gravitational force F_g between the proton m_p and the electron m_e must remain constant (see Chapter 13):

$$\frac{F_e}{F_g} = \frac{\frac{e^2}{4\pi\,\varepsilon_0 r^2}}{\frac{G\,m_p\,m_e}{r^2}} = \frac{\frac{e^2}{4\pi\,\varepsilon_0}}{G\,m_p m_e} = const. \qquad (158)$$

Since all masses fall, the masses of the proton and electron must also fall according to equations (153) and (154). Since the elementary charge$e$ and the electric field constant ε_0 are constant, this implies that the gravitational value G must increase. However, the following term remains constant:

$$G\,m_p m_e = const. \qquad (159)$$

All masses must fall at the same rate over time, so the following generally holds:

$$G\,m^2 = const. \qquad (160)$$

Due to the validity of equations (153), (154), and (160), the evolution of the gravitational value over time can now be written, as already mentioned in Chapter 13, as:

$$G_{new} = G\left(1 + \frac{\Delta t}{t_{uni\,a}}\right)^2 \qquad (161)$$

Calculation of the Earth-Moon Distance Evolution

The calculation will now be performed using all time-varying quantities according to Equation (157). For the increase in distance after one year, Δr, we obtain:

$$\Delta r = r_b - r_a \qquad (162)$$

Equation (157) can now be used to calculate the increase in distance:

$$\Delta r = \sqrt[3]{\frac{\mathrm{T}_b{}^2\,G_b(m_{1b}+m_{2b})}{4\pi^2}} - \sqrt[3]{\frac{\mathrm{T}_a{}^2\,G_a(m_{1a}+m_{2a})}{4\pi^2}} \qquad (163)$$

Substituting equations (153), (154), and (161) into equation (163) yields the following expression:

$$\Delta r = \sqrt[3]{\frac{\mathrm{T}_a{}^2\,G_a(m_{1a}+m_{2a})}{4\pi^2}\left(1 + \frac{\Delta t}{t_{uni\,a}}\right)^3} - \sqrt[3]{\frac{\mathrm{T}_a{}^2\,G_a(m_{1a}+m_{2a})}{4\pi^2}} \qquad (164)$$

Simplifying (164) yields the following calculation formula:

$$\Delta r = \sqrt[3]{\frac{T_a{}^2\, G_a(m_{1a}+m_{2a})}{4\pi^2}}\frac{\Delta t}{t_{uni\,a}} \tag{165}$$

Using the known quantities from equation (165), one arrives at the increase in distance $\Delta r = 2.8\ cm/a$ in accordance with equation (156). Currently, a distance increase of $\Delta r = 3.8\ cm/a$ is being measured using special reflectors.[31] The actual distance may deviate slightly from this value due to limited measurement accuracy.

The influence of tidal friction on the evolution of the Earth-Moon distance

Tidal friction causes the Earth's rotational energy to decrease. The braking torque generated on the Earth in this process transfers angular momentum from the Earth's rotation L_1 around its own axis with a period of T_1 to the Moon's orbital angular momentum L_2 with a period of T_2. As a result, the day is expected to lengthen by $16.5\ \mu s/a$ and the Earth-Moon distance to increase accordingly by $\Delta r = 3.8\ cm/a$.

The total angular momentum of the Earth-Moon system must be conserved. The angular momentum of the Moon's rotation about its own axis, which is coupled to its orbital motion, is comparatively small and can be neglected in this analysis. The current total angular momentum must correspond to that in one year:

$$L_1 + L_2 = const. \qquad\qquad L_{1a} + L_{2a} = L_{1b} + L_{2b} \tag{166}$$

To calculate the Earth's angular momentum, the Earth is considered an isotropic sphere with radius r_1. To account for the different mass distribution in the mantle and core in the moment of inertia, a reduced mass $m_{1red} = 0.83\ m_1$ is used: $L = 5.86\ 10^{33} kg\ m^2/s$.[32] Equation (166) leads to the following expression using the angular momenta:

$$\frac{2}{5} m_{1a\,red}\, r_{1a}{}^2 \frac{2\pi}{T_{1a}} + m_{2a} v_2 r_a = \frac{2}{5} m_{1b\,red}\, r_{1b}{}^2 \frac{2\pi}{T_{1b}} + m_{2b} v_2 r_b \tag{167}$$

The orbital velocity of the moon $v_2 = \frac{2\pi\, r}{T}\frac{m_1}{m_1+m_2}$ remains constant. The effects of (153), (154), (155), and (156) cancel each other out. Equation (167) is rearranged according to r_b, and the difference from r_a is calculated to compare the result with (156) and (165):

[31] Jürgen Müller, Why We Measure the Distance to the Moon, Yearbook of the Braunschweig Scientific Society, Leibniz University Hannover, Institute of Geodesy, 2018, p. 24

[32] Parameters of Earth Rotation and Global Dynamic Systems, Federal Agency for Cartography and Geodesy, Frankfurt am Main, 2008

$$\Delta r = r_b - r_a = \frac{\frac{2}{5}m_{1a\,red}\,r_{1a}^2\frac{2\pi}{T_{1a}}+m_{2a}v_2r_a - \frac{2}{5}m_{1b\,red}\,r_{1b}^2\frac{2\pi}{T_{1b}}}{m_{2b}v_2} - r_a = 2.8\;cm/a \quad (168)$$

The agreement is very good; equation (167) is therefore subsequently used to determine the proportion of the increase in distance due to tidal friction.

To do this, we first determine the growth of the Earth's rotation period $\Delta T_1 = T_{1b} - T_{1a}$, which, according to equation (155), would occur after one year in the absence of tidal friction:

$$\Delta T_1 = T_{1b} - T_{1a} = \mathrm{T}_{1a}\left(1 + \frac{\Delta t}{t_{uni\,a}}\right) - \mathrm{T}_{1a} = \mathrm{T}_{1a}\frac{\Delta t}{t_{uni\,a}} = 6.3\;10^{-6}\;s/a \qquad (169)$$

The observed annual lengthening of a day is $16.5\;\mu s$.[33] The, i.e., the proportion of tidal friction that slows the Earth and increases the Earth-Moon distance, is $\Delta T_1 = 16.5\;\mu s - 6.3\;\mu s = 10.2\;\mu s\;(T_{1b} = T_{1a} + 10.2\;\mu s\;)$. If equation (168) is now used such that the masses m_1, m_2, the Earth's radius r_1, and the Moon's orbital period T_2 around the Earth remain constant, the result is the pure tidal influence on the increase in distance. It amounts to:

$$\Delta r = r_b - r_a = \frac{\frac{2}{5}m_{1\,red}\,r_1^2\frac{2\pi}{T_{1a}}+m_{2a}v_2r_a - \frac{2}{5}m_{1\,red}\,r_1^2\frac{2\pi}{T_{1b}}}{m_{2a}v_2} - r_a = 0.9\;cm/a \qquad (170)$$

The influence of tidal friction is therefore much smaller than is currently assumed. Even if one attributes the entire daily elongation from (170) to the increase in the Earth-Moon distance $16.5\;\mu s$, the result is only $1.5\;cm/a$. This is not even half of the observed value, and therefore the calculation using masses that remain constant over time is questionable.

The total increase in the Earth-Moon distance calculated from equations (168) and (170) is: $\Delta r = (2.8 + 0.9)\frac{cm}{a} = \mathbf{3.7\;cm/a}$. This value is very close to the measured value of $3.8\;cm/a$ and is plausible.

The age of the universe and the influence of other celestial bodies on the Earth-Moon two-mass system, e.g., the Sun, can affect the actual distance and lead to deviations from the calculations based on equations (156), (165), (168), and (170).

Binary star systems, which are very common in the universe, would have to behave in exactly the same way. Both stars move away from each other.

---

[33] Walter Greiner, Theoretical Physics, Classical Mechanics I, Wissenschaftlicher Verlag Harri Deutsch, Frankfurt am Main, 2008, ISBN 978-3-8171-1815-1, p. 280

The view in cosmology that binary stars collide with each other while emitting gravitational waves is reversed: due to the reduction in mass, the two stars of a binary star system move away from each other. [34]

The calculated increase in the Earth-Moon distance agrees very well with observations. The expansion of space would have to cause a global mass reduction and an increase in all distances, and thus also in all scales.

If Equation (156) is used to determine the increase in the Earth's radius, one arrives at a result by Pascual Jordan. He described an annual growth of $\Delta r_1 = 0.5\ mm$. [35]

[34] Andreas Gimsa, The Expansion Dance of Alpha Centauri, International Journal of Scientific Research and Management, Volume 09, Issue 04, April 2021, DOI: 10.18535/ijsrm/v9i04.aa01
[35] Pascal Jordan, The Expansion of the Earth, Vieweg-Verlag Braunschweig, 1966, p. 77

15. Mass as the Ratio of Potential to Actualized Information

In this chapter, mass is to be defined as a purely information-theoretical concept. More than 10 years ago, I discovered the following relationship for describing mass: [36]

$$m = -\frac{1}{c^2}\frac{dm}{dt}\frac{dA}{dt} \tag{171}$$

Due to the validity of Equation (068), Equation (171) is supplemented by 4π :

$$m = -\frac{1}{4\pi c^2}\frac{dm}{dt}\frac{dA}{dt} \quad \text{or} \quad m + \frac{1}{4\pi c^2}\frac{dm}{dt}\frac{dA}{dt} = 0 \tag{172}$$

Equation (172) is a differential equation that needs to be solved. First, equation (068) is substituted into (172):

$$m = -\frac{4\pi\, tc^2}{4\pi c^2}\frac{dm}{dt} \quad \text{or} \quad \frac{m}{dm} = -\frac{t}{dt} \quad \text{or} \quad \frac{dm}{m} = -\frac{dt}{t} \tag{173}$$

Equation (173) can now be integrated for the mass$m$ from its state m_0 to its state m_1:

$$\int_{m_0}^{m_1}\frac{dm}{m} = -\int_{t_0}^{t_1}\frac{dt}{t} \qquad ln\, m\,\Big|_{m_0}^{m_1} = -\,ln\, t\,\Big|_{t_0}^{t_1} \tag{174}$$

Substituting the integration limits yields:

$$ln\, m_1 - ln\, m_0 = ln\, t_0 - ln\, t_1 \quad \text{or} \quad \frac{m_1}{m_0} = \frac{t_0}{t_1} \tag{175}$$

$$m_0\, t_0 = m_1\, t_1 \qquad \text{(see Equation (060))} \tag{176}$$

This means that the product of the mass m in the state m_0 and the time interval t_0 is exactly as large as the product of the same mass m in the state m_1 and the time interval t_1. The so-called time dilation applies to high speeds and to strong gravitational fields to which the mass m is exposed: The mass becomes larger and the time intervals become smaller. That is, time passes more slowly relative to the rest frame. During the expansion of the universe, the opposite occurs according to the same principle: Over time, the mass becomes smaller, and all time intervals on the mass become longer. This means that a present time interval on the mass passes more quickly compared to a past time interval. Isn't that fascinating? *This could influence our perception of* ***time*****: *We feel that time is passing faster and faster.*** According to Equation (068b), a longer time span has a greater information content. This relationship already appeared in the form of Equation (060) and was also verified in Chapter 12 with regard to gravity.

---

[36] Andreas Gimsa, The Beauty of Nature, 2nd edition, Gieselmann Druck und Medienhaus, Potsdam, 2014, ISBN 978-3-923830-97-8, p. 97

The relationship is universal and applies to the evolution of every mass with the age of the universe as it expands. Every mass becomes smaller as the universe expands.

In summary, according to equations (172) and (173), it can be stated that mass can be described by its own change, multiplied by the time that this change lasts.

Next, mass will be described in terms of information theory. If one divides equation (060) for the entropy of the universe, which represents its potential information, by the updated information of the universe using equation (068), the following result is obtained:

$$\frac{S_{uni}}{I_{akt\,uni}} = \frac{\frac{k}{\hbar} m_{uni}\, t_{uni}\, c^2}{4\pi\, t_{uni}\, c^2} = \frac{k}{4\pi\,\hbar}\, m_{uni} \tag{177}$$

Rearranging for mass yields a concise representation of ***mass as the ratio of potential to actualized information***.[37] The coefficient $4\pi\,\hbar/k$ is merely a constant.

$$m = \frac{4\pi\,\hbar}{k} \frac{S}{I_{akt}} \tag{178}$$

Since the potential information of the universe is constant and the updated information, according to Equation (068), increases with the expansion of the universe, the mass of the universe—and with it, the masses of all objects—must decrease. As already described in Chapter 10, the future information (see Equation (081)), which can still be updated during the expansion of the universe, consequently becomes smaller.

As explained in Chapter 9, the potential information of the state space $W = e^{S/k}$ is understood as the sum of all microstates as addressable possibilities. Thus, the quotient $N = S/k$ is the address length in Nats. Strictly speaking, it indicates the number of Nat information units required to address a microstate. N is referred to here as the number of address Nats. **Entropy is thus not understood here as disorder, but as the potential information required to address a microstate.** It is evident that W and S can be easily converted into one another, and thus both quantities can be interpreted as potential information.

In this sense, mass is the quotient of potential and actual information. It indicates the ratio of the potential address length to the address length already achieved.

---

[37] Andreas Gimsa, The Metric of Spacetime, 2nd edition, Book on Demand GmbH, Hamburg, 2020, ISBN 978-3-00-064784-0, p. 58

We see in the magnitude of the mass of the universe how far the expansion has progressed (see also the temporal evolution of the universe in Table 2). Over time, a growing address length unlocks state space that has already been created but was previously unaddressable.

Mass and Time in Special Relativity (SRT)

The validity of Equation (176) allows us to refute the special theory of relativity. Undoubtedly, for any object whose mass has been increased due to its velocity, compared to the same object at rest, time intervals on the object become shorter and clocks carried along with it run slower. If one compares the clocks of two objects moving uniformly relative to each other in two inertial frames, only the clock of the object with the higher velocity can be slower. This is because it had to be accelerated to this higher velocity through the input of energy. This is independent of the observer's point of view. The logic and Equation (176) preclude the view, permitted in SRT, that either of the two clocks can run slower relative to the other. According to Einstein, the temporal assessment according to equations (181) and (182) should apply reciprocally. This is mathematically untenable, since rearranging equation (181) does not lead to (182). At best, this would be conceivable as an observational phenomenon for massless objects, which, as is well known, do not exist in nature.

In the following scenario, two objects should have the same mass and be located in close proximity to one another. No other objects should be present in their vicinity, nor should they exert any influence on the two objects. The state of rest is denoted by the index a , and the state of motion at constant velocity is denoted by the index b .

For both masses, according to (176):

$$m_{1a}t_{1a} = m_{1b}t_{1b} \qquad\qquad m_{2a}t_{2a} = m_{2b}t_{2b} \tag{179}$$

The first mass m_{1a} is to be at rest in its inertial frame. The second mass m_2 is accelerated to the velocity$v$ and then assigned to the second inertial frame. The following relativistic formulas apply to both masses:

$$m_{1b} = \frac{m_{1a}}{\sqrt{1-\frac{v^2}{c^2}}} = \frac{m_{1a}}{\sqrt{1-\frac{0}{c^2}}} = m_{1a} \qquad\qquad m_{2b} = \frac{m_{2a}}{\sqrt{1-\frac{v^2}{c^2}}} \tag{180}$$

As a result, the moving mass of the two becomes larger at that velocity.

"A clock runs slower at a certain location" means that all time intervals are shortened there. A time interval that elapses on the moving mass is shorter than on its rest mass.

For example, if 10 minutes have elapsed on the rest mass, less than 10 minutes are guaranteed to have elapsed on the same mass moving at a constant velocity.

The Special Theory of Relativity asserts that each of the two observers sees the other's clock running slower by the Lorentz contraction factor $f = \sqrt{1 - v^2/c^2}$ than their own. [38] Even though this formulation might suggest an effect that arises solely from observation, modern science generally assumes that time actually behaves this way. According to SRT, the following should therefore hold:

$$t_{1a} = t_{2a}\sqrt{1 - \frac{v^2}{c^2}} = t_{2b} \quad (181) \qquad\qquad t_{2a} = t_{1a}\sqrt{1 - \frac{v^2}{c^2}} = t_{1b} \quad (182)$$

Each observer relates the time interval he observes in the other to his own time interval. Accordingly, the first observer, with rest time interval t_{1a}, sees the reference time interval t_{2a} of the second observer shortened by $f = \sqrt{1 - v^2/c^2}$, and the second observer, with rest time interval t_{2a}, sees the reference time interval t_{1a} of the first observer shortened by $f = \sqrt{1 - v^2/c^2}$. This is precisely what equations (181) and (182) express.

If we rearrange equation (182) in terms of t_{1a}, we obtain:

$$t_{1a} = \frac{t_{2a}}{\sqrt{1 - \frac{v^2}{c^2}}} \qquad ? \qquad (183)$$

Equation (183) naturally contradicts (181). Which of the two equations, (181) or (183), is correct? Or neither of them?

If we rearrange equation (181) according to t_{2a}, we get:

$$t_{2a} = \frac{t_{1a}}{\sqrt{1 - \frac{v^2}{c^2}}} \qquad ? \qquad (184)$$

Equation (184) contradicts (182). Which of the two equations, (182) or (184), is correct here? Or neither?

Equations (176) and (179) show that the product of mass and a time interval elapsed for the mass is constant. If equation (179) is applied to the stationary first object, with the velocity $v = 0$, only the following can hold (see also equations (008a) and (008b)):

[38] Dieter Meschede, Gerthsen Physics, 21st edition, Springer Verlag Berlin, 2002, ISBN 3-540-42024-X, p. 902

$$m_{1a}t_{1a} = m_{1b}t_{1b} = \frac{m_{1a}}{\sqrt{1-\frac{0}{c^2}}} t_{1a}\sqrt{1-\frac{0}{c^2}} \tag{185}$$

If equation (179) is applied to the moving second object, with the velocity v, only the following relationship can be mathematically correct:

$$m_{2a}t_{2a} = m_{2b}t_{2b} = \frac{m_{2a}}{\sqrt{1-\frac{v^2}{c^2}}} t_{2a}\sqrt{1-\frac{v^2}{c^2}} \tag{186}$$

While in equations (181) and (182) the parameters of both objects are mixed in an invalid manner, equations (185) and (186) satisfy the repeatedly verified equation (179).

Conclusion: Assuming the validity of equation (179), the special theory of relativity proves to be false. Although a velocity-dependent mass and a velocity-dependent time dilation are fundamentally correct, the consideration of inertial systems with massless objects is invalid.

Equation (179) refers to the state of each object independently of the other. If the masses of two objects happen to be equal and both are brought to the same constant velocity, the time intervals on both objects shorten equally in the moving state. This occurs, as a physical necessity, independently of the positions of observers.

Only a mass that has been accelerated to a certain velocity has more energy than it does at rest. Therefore, every interval of time is shortened for a mass moving at high speed. The positions of any observers are irrelevant to the passage of time for the objects.

When an observer on a uniformly moving object measures time, their clock runs slower there compared to the same object at rest. This is consistent with observations conducted using airplanes. [39]

An example of relativity: We are sitting in the direction of travel on a train that is standing at the platform. On the adjacent track, a train of the same mass is departing in the opposite direction. Looking out, we might get the impression that we are moving ourselves. A little later, the other train reaches its constant final velocity; we are still standing still.

[39] Hafele, Keating, “Around the World with Atomic Clocks: Predicted Relativistic Time Gains,” Science 177, No. 4044, 1972, pp. 166–168

Even though it makes no difference to the increasing distance between the trains which train has departed, provided they are accelerated equally and also reach the same final velocity, there is a crucial difference: only on the train to which energy was supplied to reach the final velocity do the passengers' clocks run slower and their time intervals become shorter. This fundamental law is enshrined in Equation (179). Only when both trains are accelerated to the same speed in opposite directions do the same time-related conditions arise on each train.

The passengers on the stationary train may rightly claim that the clocks of the passengers on the moving train run slower than their own. The reverse, however, is not true. For in reality, the clocks of the passengers on the stationary train do not run slower, but conversely, faster than those on the moving train.

The relativity of perception does not, therefore, replace the physical relativity of existence, which places every energetic state within a specific temporal context.

Conclusion: The equations of the Special Theory of Relativity, expressed by formulas (181) and (182) and intended to describe time dilation, are simply incorrect. There are neither massless objects nor a temporally equivalent behavior of objects with different energy states.

16. Dark Matter: an attempt at an explanation

It has been observed that the velocity of stars rotating in the outer regions of galaxies around the galactic center is higher than expected. The velocities tend toward a constant value as they move outward. Therefore, it is suspected that dark matter is present in galaxies, which would have to be added to the galactic mass $M(r)$. This would explain the deviations in the velocities, whose expected values follow from the known formulae.

Based on the balance between centripetal and gravitational forces, it is expected that the velocities of stars in the outer orbits of galaxies decrease with their distance r from the galactic center:

$$F_Z = F_{grav} = m\frac{v^2}{r} = G\frac{M(r)\,m}{r^2}$$

Here, m is the mass of the orbiting star and $M(r)$ is the mass of the galactic center.

$$v_{star} = \sqrt{G\frac{M(r)}{r}} \qquad v_{star} \sim \sqrt{\frac{1}{r}} \tag{187}$$

For outwardly, with r, the enclosed mass region $M(r)$ —whose gravitational pull acts on an outwardly rotating star—becomes only slightly larger. It can be regarded as constant.

For the universe as a whole, the following applies to its baryonic and dark matter:

$$M_{bm} = \frac{t_{uni}\,c^3}{G} = \frac{r_{uni}\,c^2}{G} \tag{188}$$

$$M_{dm} = \frac{t_{i\,uni}\,c_i{}^3}{G} = \frac{r_{uni}\,c^2}{G} \tag{189}$$

Here, the dark mass M_{dm} is defined using the imaginary time t_i. If, instead of the speed of light c , we substitute the orbital velocity v of stars at a distance$r$ from the galactic center into equations (188) and (189), we arrive at the equation form (187):

$$M_{bm} = \frac{r\,v_{bm}{}^2}{G} \tag{188a}$$

$$M_{dm} = \frac{r\,v_{dm}{}^2}{G} \tag{189a}$$

This assumes that dark matter exists not only in the universe but also in every galaxy.

As equation (189) shows, the dark mass M_{dm} is positive, just like the baryonic mass. Thus, it also has an attractive effect, since the imaginary unit i is positive to the fourth power $i^4 = \sqrt{-1}^4 = 1$.

Here it is evident that a portion of the positive mass of the universe could exist as dark mass.

If we combine the velocity components of both types of mass for a star orbiting within the galaxy, we obtain the following expression:

$$v_{star} = v_{bm} + v_{dm} = \sqrt{G\frac{M_{bm}(r)+M_{dm}(r)}{r}} \quad (190)$$

Due to the dark mass M_{dm}, the velocity v is higher than expected because of a larger total mass $M_{bm}(r) + M_{dm}(r)$ compared to the baryonic mass $M_{bm}(r)$.

Dark matter could therefore arise from the effect of the ether in the interaction of imaginary time with imaginary light. What applies to the universe could also apply to every galaxy: In the outer regions of the galaxy, the stars rotate faster than expected. This unexpected balance of centripetal and gravitational forces at higher stellar velocities can be easily explained by equation (190).

The indirect observation of dark matter through the rotational velocity of stars would thus also serve as evidence for the existence of imaginary light.

Dark matter can thus also serve as evidence for the existence of imaginary time in Equation (189), which was described as the missing element in the description of the universe (see Chapters 2 and 5).

What defines our gut feeling? It is clearly information that should be taken seriously. Information that works well without any rational explanation or lengthy deliberation. A survival instinct, and we often only realize later why it was good and right for us. It is mathematically imaginary information that we access through imaginary time. The spirituality that nourishes our soul and allows us to bypass rational thinking lets us dive deep into imaginary time. Instantaneous events in different places and natural quantum entanglements likely open up the horizon for us to benefit from the experience of the entire universe. We do not need to travel to distant places to experience this knowledge: Imaginary time creates a gateway for us to the chronicle of the universe. The universe knows the truth and cannot be manipulated. Present and past information from all locations is, in principle, available. We need only look deep into space to recognize events from the past.

17. The Structure of the Proton [40]

This chapter deals with the structure of the proton. According to these new findings, the proton consists of three main elements: a magnetic north pole, a magnetic south pole, and a positron. What are the implications of this structure, and how can it be proven?

The previous description of the proton involving quarks and gluons may be incorrect. My hypothesis that magnetic north and south poles can form electron neutrinos independently of the proton has also been rejected.

The positron would be responsible for the proton's positive electric charge. For example, a special exotic form of hydrogen known as positronium is known, whose nucleus contains not a proton but a positron. Due to the absence of the proton, positronium is much lighter than normal hydrogen and, together with its electron, is also electrically neutral overall.

The atomic structure of matter in the universe, which consists primarily of hydrogen, would be extraordinarily symmetrical and beautiful:

There would not only be the same number of positive and negative elementary electric charges, but also the same number of magnetic north and magnetic south poles, and thus magnetic elementary charges in the universe. The total number of elementary electric and magnetic charges would be identical. With the positron, electron, north pole, and south pole, the simplest and most common hydrogen atom in the universe comprises 4 fundamental elements, 2 of which are electric elementary charges and 2 of which are magnetic elementary charges. If the magnetic south pole were the antiparticle of the magnetic north pole—an anti-north pole—then the same number of material and antimatter elementary particles would exist. The sought-after magnetic elementary charges and the equally sought-after antimatter would thus be found.

In the publication, [41] both the number of electric charges N and that of the magnetic charges N in each mass were calculated using the simple relationship $N = m/\left(m_p + m_e\right)$. This is also a condition for the correct calculation of gravitational force, as discussed below. From this, and based on the structure of the proton, it was concluded that the same applies to the number of magnetic poles: $N = m/(m_{NP} + m_{SP})$. Since both magnetic poles are components of the proton, this is readily possible.

[40] Andreas Gimsa, The Structure of the Proton and the Combination of Fundamental Forces, International Journal of Scientific Research and Management, Volume 01, Issue 10, January 2024, DOI: 10.18535/ijsrm/v12i01.p01

[41] Andreas Gimsa, The electromagnetic cause of gravity, International Journal of Scientific Research and Management, Volume 11, Issue 10, Oct. 2023, DOI: 10.18535/ijsrm/v11i10.p01

The north pole m_{NP} and the south pole m_{NP} must have the same mass : $m_{NP} = m_{SP}$. How can it be explained that the magnetic north poles and their possible antiparticles, the south poles (anti-north poles), do not annihilate each other? It was shown in my earlier publications that magnetic monopoles could be involved in the formation of time.[42] The anti-north pole and the positron could play a special role in this. It is probably not readily possible for both magnetic poles to meet at the same time in one place. By law of nature, they presumably have a minimal temporal separation. The sources and sinks of the elementary magnetic charges could be located within the masses and form time.

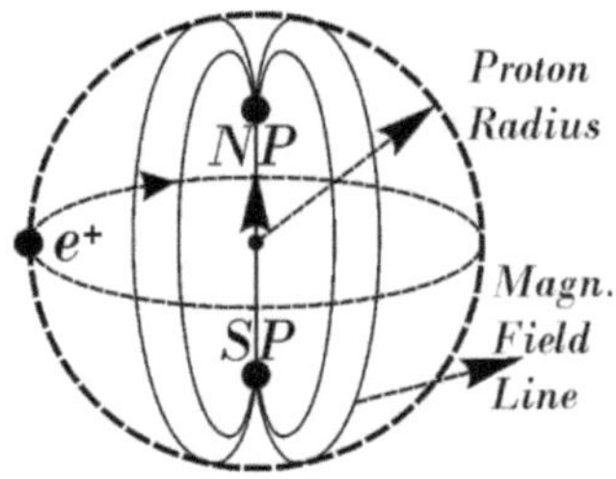

Figure 9: *Model of the proton, consisting of a north pole (NP), a south pole (SP), and a positron (e+)*

Why do the elementary electric particles, the electron and the positron, not annihilate each other? The electric positron is shielded from the electron and kept at a distance by binding to magnetic monopoles in the proton. Only theoretically is it possible, with the help of the tunnel effect, for both an annihilation of the electron and the positron and an annihilation of the north pole and south pole to occur.

The Maxwell equations would become perfectly symmetric with the two elementary magnetic charges, which would then represent the sources and sinks of magnetic fields. An important observational gap could be closed. Dirac had already suspected that elementary magnetic charges must exist in nature; they simply had not been discovered yet.

With the help of the three charges involved, the proton's great stability and its magnetic moment could be explained.

The fact that a proton fundamentally consists of three elements is supported by observation. It should be noted here that different combinations of the three elements lead to various elementary particles. These would now need to be re-explained with reference to the newly discovered three elements of the proton.

[42] Andreas Gimsa, Magnetic Monopoles, Gravity, and Time: Derivation from Maxwell's Equations, Kuss GmbH, Potsdam, Germany, 2016, ISBN 978-3-00-053660-1

This concerns their structure and their interactions. It is currently assumed that certain elementary particles are composed of quarks and thus of thirds of an elementary charge. However, this contradicts the observation of integer charges and should be rejected for that reason alone. Magnetic charges, in interaction with the positron and the acting Lorentz force, could create the false impression that they are electric charges in certain parts smaller than an elementary charge.

For example, when one object is attracted to another permanent magnetic object, it is not immediately apparent from the external effect which of the two interacting objects is the magnet, whether both objects are magnets, or whether both objects simply possess opposite electrostatic charges.

The Forces Within the Proton and Their Evolution

We will now examine the forces acting within the proton. These forces ensure the proton's cohesion. The Lorentz force between the north pole and the positron, as well as between the south pole and the positron, could be greater than the magnetic force between the north and south poles by a ratio of strong force to magnetic force. In that case, the strong force could simply be the resulting Lorentz force that arises from the configuration of the charges and their relative motions within the proton. This force also acts beyond the boundaries of the proton and ensures the cohesion of protons with one another.

Between the north and south poles, at a distance of r, the following attractive magnetic force acts:

$$F_{mg} = \frac{p^2}{4\pi\,\mu_0\,r^2} \tag{191}$$

This relationship is comparable to that for the force between two electric charges, for example, the proton and the electron:

$$F_{el} = \frac{e^2}{4\pi\,\varepsilon_0\,r^2} \tag{192}$$

The distance r according to Equation (191) has the same order of magnitude as the proton radius r_p within the proton.

In the following discussion, we assume that the two magnetic poles and each magnetic pole are at the same average distance from the electric pole. Since the location of a quantum must be described in terms of its probability of being there, this is a probable distance.

One can visualize this as an equilateral triangle in which a magnetic pole is located at each of two corners and the positron is located at one corner (see Fig. 9).

The magnetic force according to Equation (191) is $F_{mg} = 3.260\ 10^2\ N$, with the proton radius $r_p = 8.412\ 10^{-16}\ m$ as the distance and the elementary magnetic charge $p = 6.036\ 10^{-17}\ Vs$.

The Lorentz force between a magnetic pole and the positron inside the proton is now to be quantified. The known Lorentz force acting on a moving electric charge $\vec{Q}_{el}$, which moves at a speed of $\vec{v}$ in an electromagnetic field, is calculated as follows:

$$\vec{F}_{Lorentz} = \vec{Q}_{el}\vec{E} + \vec{Q}_{el}(\vec{v} \times \vec{B}_{mg}) \tag{193}$$

If no external electric field of field strength $\vec{E}$ is present, equation (193) simplifies, and the resulting force $\vec{F}_{res}$ acts on the electric charge $\vec{Q}_{el}$:

$$\vec{F}_{res} = \vec{Q}_{el}(\vec{v} \times \vec{B}_{mg}) \tag{194}$$

If the velocity of the electric charge $\vec{Q}_{el}$ is perpendicular to the field lines of the magnetic field with flux density $\vec{B}_{mg}$, equation (04) can be further simplified by omitting the cross product:

$$\vec{F}_{res} = \vec{Q}_{el}\, \vec{v}\, \vec{B}_{mg} \tag{195}$$

In the magnetic field with flux density $\vec{B}$ emanating from one of the two magnetic monopoles, the resulting force $\vec{F}_{res}$ acts on the positron moving at velocity $\vec{v}$.

The charge of the positron corresponds to the elementary electric charge:

$$\vec{Q}_{el} = e^- = e^+ = e \tag{196}$$

In the classical model, the positron's velocity $\vec{v}_{po}$ could be nearly equal to the speed of light. The velocity is perpendicular to the magnetic field axis between the north and south poles. It is the limiting velocity for massive particles, which the positron thus nearly reaches.

$$\vec{v}_{po} \approx c \tag{197}$$

This is not a classical velocity in space. This velocity should be understood as a measure of the relativistic energy that the quantum would possess if it were moving through space at this velocity.

It is assumed that each quantum in the proton, with its relativistic energy, accounts for one-quarter of the proton's rest energy. In a symmetrical arrangement, the rest energy of the positron and the two magnetic monopoles could correspond to the rest energy of the electron. Since the proton has only three components, the remaining quarter of the proton's mass could be electromagnetic radiation that can be emitted in the form of time.

Like the fine-structure constant, which can be interpreted as the ratio of the electron's velocity to the speed of light in the hydrogen atom at the lowest energy level, this is a conceptual model. At 99.99976% of the speed of light, the positron is about 459 times heavier than in its rest state. The relativistic mass of the positron could thus contribute a significant portion to the proton mass. The positron mass is examined in greater detail below. In this model, the electron, which lies much farther out in the hydrogen atom, has a velocity relative to the positron that is approximately 1/137 of the speed of light.

The arrangement of the north pole and south pole in the proton can be understood as a magnetic dipole, whose two poles have a probable spatial (and temporal) distance between them.

Figure 1 shows the model arrangement of the orbital radii of the three poles. From a quantum-physical perspective, this classical description is untenable. The quanta must be described using a wave function. Since the three quanta in the proton are likely in a mixed and/or entangled state, the description is only possible via a quantum mechanical density matrix. This approach, which agrees with observations for other processes, has been confirmed many times and yet is not fully understood.

The fundamental configuration of the proton's structure, with one electric and two magnetic poles, is particularly promising because of the symmetry it achieves, which fits well with the description of nature.

The uncertainty of the locations of quanta could be explained in part by the formation of time via antiparticles (Equations (081a) and (081b)). This results in both the location of the quantum at a specific point in time and its velocity-dependent momentum no longer being classically calculable. For a new time has been added that would have to be taken into account. Due to the formation of time, which represents an update of information, the quantum becomes slightly less energetic (see Chapter 10).

In my opinion, the description of uncertainty includes the imaginary second temporal dimension into which the quantum can presumably evade instantaneously (at imaginary light speed).

In this way, the quanta in the proton and also the electron in the hydrogen atom strive toward their lowest-energy ground state. This is achieved when the quantum emits no synchrotron radiation, as is otherwise common with accelerated charges. If the quantum is completely in the imaginary present, it is timeless and untraceable.

The electron is attracted to the positron in the proton and could, in principle, annihilate with it. Similarly, the magnetic north pole could annihilate with its antiparticle, the magnetic south pole. How does nature prevent this from happening? One possibility would be a defined quantum entanglement that acts in such a way that, despite the minimal quantum motion, either a constant temporal or a constant spatial distance is maintained: If one quantum moves toward the other, the other quantum adjusts its temporal or spatial distance to maintain the separation. They can then only annihilate if the entanglement is broken.

According to Equation (197), the velocity of the positron is $2.998\ 10^8\ m/s$.

The magnetic flux density $\vec{B}$ in Equation (195) is derived from the magnetic field strength. For a point-like magnetic charge in space, the magnetic field strength $\vec{H}$ at a distance r is given by:

$$\vec{H} = \frac{\vec{Q}_{mg}}{4\pi\mu_0\, r^2} \tag{198}$$

This yields the magnetic flux density $\vec{B}$ with the absolute permeability μ and the magnetic elementary charge p:

$$\vec{B} = \mu\vec{H} = \mu_0\mu_r\vec{H} = \mu_0\mu_r\frac{p}{4\pi\mu_0\, {r_p}^2} = \mu_r\frac{p}{4\pi\, {r_p}^2} \tag{199}$$

The material-dependent permeability μ_r within the proton is the ratio of the Klitzing resistance R_{kl} to the vacuum wave resistance Z_0:

$$\mu_r = \frac{R_{kl}}{Z_0} \tag{200}$$

The magnetic flux density according to Equation (199) can now be calculated. It is given by (200): $\vec{B} = 5.401\ 10^{18}\ Vs/m^2$. The resulting electromagnetic Lorentz force on the positron according to Equation (195) in the magnetic field with flux density $\vec{B}$ can then be written as follows:

$$\vec{F}_{res} = \vec{Q}_{el}\,\vec{v}\,\vec{B} = \frac{epc}{4\pi\, {r_p}^2}\,\mu_r = \frac{epc}{4\pi\, {r_p}^2}\,\frac{R_{kl}}{Z_0} \tag{201}$$

Here, the dimensionless relationship I have already published applies: [43]

$$\frac{R_{kl}}{Z_0} = \frac{1}{2\,\alpha_{em}} = \frac{\alpha_{strong}}{\alpha_{em}} \tag{202}$$

This allows us to simplify equation (201) and calculate the force:

$$\vec{F}_{res} = \frac{epc}{8\pi\,\alpha_{em}\,r_p{}^2} = \frac{R_{kl}}{Z_0}F_{mg} = \mu_r F_{mg} = 2.234\;10^4\;N \tag{203}$$

Thus, according to Equation (203), the electromagnetic Lorentz force—as a strong force—is 68.5 times greater than the magnetic force given by Equation (191).

The quark-gluon model is thus falsified. It is noteworthy that, due to the permeability in the proton—which differs from that in a vacuum—and the speed of the positron, which is nearly equal to the speed of light, the Lorentz force is exactly as much greater as the permeability in the proton compared to the comparable magnetic force in a vacuum. The structure of the proton shows that the strong force is a Lorentz force (see also Fig. 13). It arises from the electromagnetic interaction between the two magnetic poles and the positron within the proton.

The unique structure of the proton and the resulting perfect symmetry of the hydrogen atom and the universe undoubtedly represent one of the greatest wonders of nature.

The material-dependent permeability μ_r in the proton according to (200) can also be written as the ratio of the strong force to the electromagnetic force:

$$\mu_r = \frac{F_{strong}}{F_{em}} = \frac{\alpha_{strong}}{\alpha_{em}} = \frac{R_{kl}}{Z_0} = 68.5 \tag{204}$$

In other words, the strong force is an electromagnetic force that is greater in the proton because the permeability there is greater than in a vacuum.

Let us consider the energy that would be released if two monopoles in the proton were separated by an infinite distance from their proton radius. Based on equation (201), the following applies to two magnetic monopoles in the proton:

$$E = \int_{r_p}^{\infty} F\,ds = \int_{r_p}^{\infty} \frac{\mu_r\,p^2}{4\pi\,\mu_0\,s^2}ds = -\frac{\mu_r\,p^2}{4\pi\,\mu_0\,s}\bigg|_{r_p}^{\infty} \tag{205}$$

$$E = -\left(\frac{\mu_r\,p^2}{4\pi\mu_0\,\infty} - \frac{\mu_r\,p^2}{4\pi\,\mu_0\,r_p}\right) = \frac{\mu_r\,p^2}{4\pi\mu_0\,r_p} = \frac{1}{8}m_p c^2 \tag{206}$$

The new result according to Equation (206) is already remarkable, because this energy corresponds exactly to one-eighth of the proton's rest energy.

[43] Andreas Gimsa, The electromagnetic origin of gravity, International Journal of Scientific Research and Management, Volume 11, Issue 08, August 2023, DOI: 10.18535/ijsrm/v11i08.p01

In the following, novel energy balances are established that could be derived from the structure of the proton. According to Fig. 9, an energy component must exist between two magnetic charges, a second component of equal magnitude between the magnetic north pole and the positron, and a third between the magnetic south pole and the positron. It is noteworthy that there is a fourth energy component of equal magnitude with the positron's frequency f_{po}, from which time could arise during the expansion of the universe and thus with the expansion of the proton.

$$E_p = m_p c^2 = \frac{1}{\alpha_{strong}} \left(\frac{\mu_r \, p^2}{4\pi\mu_0 \, r_p} + \frac{\mu_r \, e^+ p_{NP} \, c}{4\pi \, r_p} + \frac{\mu_r \, e^+ p_{SP} \, c}{4\pi \, r_p} + \mu_r h f_{po} \right) \tag{207}$$

Using Equation (204), Equation (207) can also be written with the fine-structure constantα for the proton energy E_p:

$$E_p = m_p c^2 = \frac{1}{\alpha_{em}} \left(\frac{p^2}{4\pi\mu_0 \, r_p} + \frac{e^+ p_{NP} \, c}{4\pi \, r_p} + \frac{e^+ p_{SP} \, c}{4\pi \, r_p} + h f_{po} \right) \tag{208}$$

It is quite astonishing that Equation (208) can be used to calculate exactly the rest energy of the proton. This is a clear indication of its structure.

Equation (208) resembles that in the Planck system, where the following applies to the energy:

$$E_{planck} = m_{planck} c^2 = \frac{1}{\alpha_{em}} \frac{p^2}{4\pi\mu_0 \, r_{planck}} = \frac{1}{\alpha_{em}} \frac{e^2}{4\pi\varepsilon_0 \, r_{planck}} \tag{209}$$

Consequently, in the Planck system, with the strong force interaction constant α_{st}, the following can also be written:

$$E_{planck} = m_{planck} c^2 = \frac{1}{\alpha_{strong}} \frac{\mu_r \, p^2}{4\pi\mu_0 \, r_{planck}} \tag{210}$$

It follows directly from this that if the distance between two magnetic monopoles is as small as the current Planck radius, the Planck rest energy is equal to the energy released when the monopoles are moved an infinite distance apart from this position. Since the same permeability applies here as in the proton, the Planck rest energy may represent a portion of the energy of the smallest proton that has ever existed. Since all distances are constantly growing, the distance between two monopoles may also have once been as small as the Planck radius.

Let us turn to the fourth term of Equation (208), from which the time that is constantly being created in the universe could radiate. This energy component amounts to one-quarter of the proton's rest energy. During the expansion of the universe, all protons grow larger. As equation (208) shows us, this causes the rest energy of the proton to decrease. The frequency f_{po} must also decrease, and accordingly, the reduced period t_{po} must increase.

This could mean that all protons radiate time as the universe expands. The same radiation process, but with the opposite sign, is described in Chapter 19 in the context of gravity. While time is created as the universe expands, it is eliminated during gravity.

$$E_{time} = \frac{E_p}{4} = \frac{1}{\alpha_{em}} h f_{po} = \frac{2\pi}{\alpha_{em}} \hbar f_{po} = \frac{2\pi}{\alpha_{em}} \frac{\hbar}{T_{po}} = \frac{1}{\alpha_{em}} \frac{\hbar}{t_{po}} \qquad (211)$$

The creation of time could arise from the antiparticle positron and the magnetic south pole of the protons, which move, so to speak, backward in time. Equations (081a) and (081b) illustrate the physical process.

The proton has already posed many puzzles for physicists. Is it worth seriously considering that the structure described here is correct?

<u>The following reasons support this:</u>

The internal dynamics of the proton reveal a seething quantum object in which only fields are discernible. Quarks and gluons are particularly well-suited as models for energy balance calculations.

No single free quark has ever been observed.

In the quark/gluon model, there are one-third charges, whereas charges in nature otherwise always occur as integers. The three valence quarks with one-third charges explain only one-third of the proton spin. According to the model presented here, there are three integer charges (Fig. 9).

The proton structure can explain the strong force as an electromagnetic force.

The proton mass is precisely explained by the charge model presented here.

The proton radius used here, which is considered established in terms of magnitude, plays a decisive role in the presented energy calculation.

The proton structure presented here yields the wonderful charge symmetry in the universe already described.

According to Einstein, it is the theory that determines what can be observed. This is a profound insight that should be applied to the proton structure presented here.

As I write this, it occurs to me that a magnetic monopole, with the described structure of the proton and its effect on gravity, is also an excellent candidate for the book title "The Missing Element." If anyone prefers it, they are welcome to choose this reference.

18. The Electromagnetic Cause of Gravity [44]

Assuming the expansion of the universe and all the masses within it, along with their atoms and charges, the electromagnetic cause of the gravitational force is determined using the Lorentz force. The gravitational force, which has been shown here to have an electromagnetic basis, is identical to Newton's mechanical gravitational force, which, although repeatedly confirmed experimentally and correct, remains unsubstantiated.

The displacement currents during the expansion of the universe

The expansion of the universe affects all masses spatially through the enlargement of their atoms. Consequently, this also affects the electron orbitals. It is assumed here that atoms contain not only electric but also magnetic elementary charges. The elementary magnetic charges cannot exist in space as free charges. They could be quanta of protons and elude direct spatial observation. This could be the reason why only closed magnetic field lines are perceived in space. The sources of the elementary magnetic charges are located in time. The elementary magnetic charges could be responsible for the strong force within the protons.

The expansion of all atoms during the expansion of the universe leads to displacement currents, since the distances between the charges contained within the atoms also increase. The displacement current is not a current in which a charge is transported. It corresponds to a change in an electric or magnetic flux.

The calculations presented here, with their correct and observable results, show that the displacement currents, which are illustrated in Figures 10 and 11, always have a defined spatial direction. The current is perpendicular to the axis of motion of two masses that fall toward each other from an assumed state of rest due to the gravitational effect. This uniform direction of the displacement current is directly related to the uniform direction of time, which is perpendicular to the three spatial dimensions and runs from the past into the future. While both masses are initially in their rest positions, they will move along their axis of motion and collide with each other in the future.

There is always a pair of displacement currents, one electric and one magnetic, each with a defined direction that is perpendicular to the other.

---

[44] Andreas Gimsa, The Electromagnetic Cause of Gravity, International Journal of Scientific Research and Management, Volume 11, Issue 10, Oct. 2023, DOI: 10.18535/ijsrm/v11i10.p01, The German-English print version can be found under ISBN 978-3-00-076845-3

The displacement current can be understood as a quantity of charge moving over time.

It is to be demonstrated that gravity is a consequence of the expansion of the universe. From the following explanations, it can be concluded that it could not occur in a static universe.

The electric displacement current $\vec{I}_{el\,1}$ (Fig. 10) is accompanied by a displacement of the charges $\vec{Q}_{el\,1}$. The magnetic flux density $\vec{B}_1$ is a consequence of the magnetic displacement current $\vec{I}_{mg\,1}$, which is perpendicular to the electric displacement current. It does not occur in space and therefore cannot be depicted in the spatial representation of Fig. 10. This magnetic displacement current inhibits time in the region of the mass m_1. Until now, physics has assumed time to be a given, yet it arises only according to electromagnetic rules with the expansion of the universe in the region of masses (see Chapter 10).

The electric displacement current $\vec{I}_{el\,2}$ (Fig. 10) is accompanied by a charge displacement of the charges $\vec{Q}_{el\,2}$. The magnetic flux density $\vec{B}_2$ is a consequence of the magnetic displacement current $\vec{I}_{mg\,2}$, which is perpendicular to the electric displacement current. It does not occur in space and therefore cannot be depicted in the spatial representation of Fig. 10. This magnetic displacement current inhibits time in the region of mass m_2.

The electric displacement current $\vec{I}_{el\,1}$, together with the magnetic flux density $\vec{B}_2$—which acts on the mass m_1 from a distance—accounts for half of the gravitational force. This process can be described using the "right-hand rule." Here, the thumb points in the direction of the electric displacement current $\vec{I}_{el\,1}$, the index finger points in the direction of the magnetic field lines of the magnetic field with flux density $\vec{B}_2$, and the middle finger points in the direction of the resulting Lorentz force.

The electric displacement current $\vec{I}_{el\,2}$, together with the magnetic flux density $\vec{B}_1$, which acts on the mass m_2 from a distance, results in the other half of the gravitational force. This process can also be described using the “right-hand rule.” Here, the thumb points in the direction of the electric displacement current $\vec{I}_{el\,2}$, the index finger points in the direction of the magnetic field lines of the magnetic field with flux density $\vec{B}_1$, and the middle finger points in the direction of the resulting Lorentz force.

The two Lorentz forces from the two displacement currents add together to form the gravitational force.

Fig. 11 depicts a process symmetrical to Fig. 10, in which the magnetic displacement currents $\vec{I}_{mg}$ do not occur invisibly in time but are observable in space. The electric field lines of the electric fields with flux densities $\vec{D}_1$ and $\vec{D}_2$ are closed and have no sources in space. This process cannot occur during the expansion of the universe, as no magnetic elementary charges are observed. If, instead of endless expansion, one assumes a pulsation of the universe, a contraction phase is conceivable in which the electric and magnetic components of matter are reversed. The examination of the process symmetrical to Fig. 10 as shown in Fig. 11 and the corresponding calculations are therefore of purely theoretical value and are presented here because it was found that exactly the same gravitational force results.

The electric and magnetic flux densities $\vec{D}$ and $\vec{B}$ shown in Figs. 10 and 11 do not result from electric or magnetic fields acting outward, but rather from the change in electric and magnetic flux due to the spatial enlargement of the atoms. The displacement current flows even in a vacuum when a field changes there. It can be expressed as the time derivative of the charge. For the electric displacement current, the relationship known from Maxwell applies:

$$\vec{I}_{el} = \varepsilon \int_A \frac{\partial \vec{E}}{\partial t} d\vec{A} = \varepsilon_0 \varepsilon_r \int_A \frac{\partial \vec{E}}{\partial t} d\vec{A} = \varepsilon_r \int_A \frac{\partial \vec{D}}{\partial t} d\vec{A} = \frac{d\vec{Q}_{el}}{dt} \quad (212)$$

Just as there is an electric displacement current, according to the full Maxwell equations, if there are magnetic elementary charges in the atoms, there must be a magnetic displacement current that flows when the charge increases:

$$\vec{I}_{mg} = \mu \int_A \frac{\partial \vec{H}}{\partial t} d\vec{A} = \mu_0 \mu_r \int_A \frac{\partial \vec{H}}{\partial t} d\vec{A} = \mu_r \int_A \frac{\partial \vec{B}}{\partial t} d\vec{A} = \frac{d\vec{Q}_{mg}}{dt} \quad (213)$$

The displacement currents at the masses m_1 and m_2 according to equations (212) and (213) are illustrated in Figures 10 and 11.

The Electromagnetic Cause of Gravity

The well-known Lorentz force acting on a moving electric charge $\vec{Q}_{el}$, which moves at a velocity $\vec{v}$ in an electromagnetic field, is calculated as follows:

$$\vec{F}_{Lorentz} = \vec{Q}_{el}\vec{E} + \vec{Q}_{el}(\vec{v} \times \vec{B}_{mg}) \quad (214)$$

If there is no external electric field with a field strength of $\vec{E}$ acting on the electric charge, equation (214) simplifies to the resulting force $\vec{F}_{res}$:

$$\vec{F}_{res} = \vec{Q}_{el}\left(\vec{v} \times \vec{B}_{mg}\right) \tag{215}$$

If, as shown in Fig. 10, the velocity of the electric charge $\vec{Q}_{el}$ in the form of the electric displacement current $\vec{I}_{el}$ is perpendicular to the field lines of the magnetic field with flux density $\vec{B}_{mg}$, equation (215) can be further simplified by omitting the cross product:

$$\vec{F}_{res} = \vec{Q}_{el}\,\vec{v}\,\vec{B}_{mg} \tag{216}$$

The mass m_1 in Fig. 10 is acted due to the electric displacement current $\vec{I}_{el\,1}$ in the magnetic field with flux density $\vec{B}_2$, which emanates from the mass m_2, the following resulting force:

$$\vec{F}_{res1} = \vec{Q}_{el\,1}\,\vec{v}_1\,\vec{B}_2 \tag{217}$$

The mass m_2 in Fig. 10 is acted due to the electric displacement current $\vec{I}_{el\,2}$ in the magnetic field with flux density $\vec{B}_1$ which emanates from the mass m_1, the following resulting force:

$$\vec{F}_{res2} = \vec{Q}_{el\,2}\,\vec{v}_2\,\vec{B}_1 \tag{218}$$

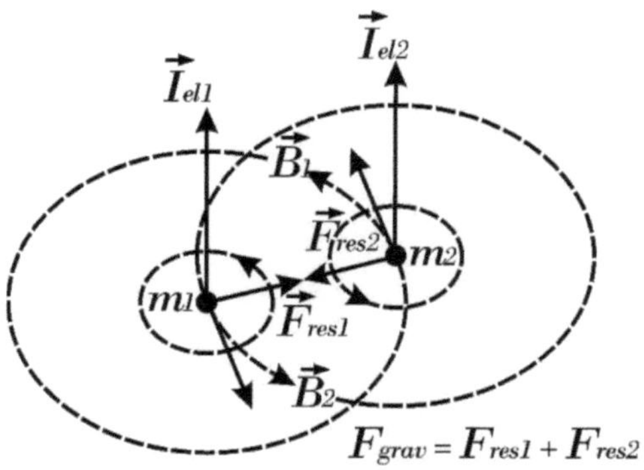

Figure 10 Origin of the gravitational force due to the electric displacement current during the expansion of the universe

If the universe pulsates and if magnetic monopoles also exist in the masses during the contraction of the universe, then based on Maxwell's equations there should be a force comparable to the Lorentz force acting on a moving magnetic charge $\vec{Q}_{mg}$ moving at a velocity $\vec{v}$ in an electromagnetic field:

$$\vec{F}_{res} = \vec{Q}_{mg}\vec{H} + \vec{Q}_{mg}\left(\vec{v} \times \vec{D}_{el}\right) \tag{219}$$

If no external magnetic field of field strength $\vec{H}$ is present for the magnetic charge, equation (219) simplifies to:

$$\vec{F}_{res} = \vec{Q}_{mg} \left(\vec{v} \times \vec{D}_{el}\right) \tag{220}$$

If, as shown in Fig. 11, the velocity of the magnetic charge $\vec{Q}_{mg}$ in the form of the magnetic displacement current $\vec{I}_{mg}$ is perpendicular to the field lines of the electric field with flux density $\vec{D}_{el}$, equation (220) can be further simplified by omitting the cross product:

$$\vec{F}_{res} = \vec{Q}_{mg}\, \vec{v}\, \vec{D}_{el} \tag{221}$$

The mass m_1 in Fig. 11 is acted due to the magnetic displacement current $\vec{I}_{mg\,1}$ in the electric field with flux density $\vec{D}_2$, which emanates from the mass m_2, the following resulting force:

$$\vec{F}_{res1} = \vec{Q}_{mg\,1}\, \vec{v}_1\, \vec{D}_2 \tag{222}$$

The mass m_2 in Fig. 11 is acted due to the magnetic displacement current $\vec{I}_{mg\,2}$ in the electric field with flux density $\vec{D}_1$ which emanates from the mass m_1, the following resulting force:

$$\vec{F}_{res2} = \vec{Q}_{mg\,2}\, \vec{v}_2\, \vec{D}_1 \tag{223}$$

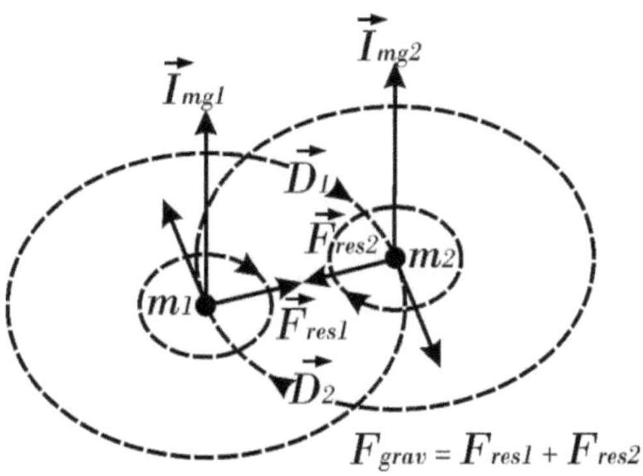

Figure 11 Origin of the gravitational force due to the magnetic displacement current during the theoretical contraction of the universe

The individual coefficients of equations (217), (218), (222), and (223) are to be determined and assigned numerical values for the **Earth-Moon** gravitational system. Calculations are performed using the average distance$r$ between their centers.

Equation (217) is used to determine the resulting force $\vec{F}_{res1}$ due to the electric displacement current $\vec{I}_{el\,1}$ in the magnetic field with flux density $\vec{B}_2$ (Fig. 10):

The total electric charge $\vec{Q}_{el\,1}$ on the mass m_1 is determined by the number of individual charges. The number of individual charges, in turn, is calculated using the object's mass m_1, the proton mass m_p, and the electron mass m_e:

$$\vec{Q}_{el\,1} = N_{e1}e \qquad \text{where} \qquad N_{e1} = \frac{m_1}{m_p+m_e} \tag{224}$$

For the Earth, with a mass of m_1, this results in $N_{e1} = 3.571\ 10^{51}$ elementary electric charges. The total electric charge according to Equation (224) is accordingly $\vec{Q}_{el\,1} = 5.722\ 10^{32}\ As$.

The velocity $\vec{v}_1$ of the displacement current $\vec{I}_{el1}$ resulting from the atomic expansion with the expansion of the universe is extremely slow and corresponds to the ratio of the Bohr radius r_{bohr} to the age of the universe t_{uni}. While the radius of the universe r_{uni} grows at the speed of light, the Bohr radius r_{bohr} grows extremely much more slowly.

$$\vec{v}_1 = \frac{r_{bohr}}{t_{uni}} \quad \text{with} \quad t_{uni} = 13.8\ 10^9 a \quad \text{and} \quad r_{bohr} = \frac{4\pi\varepsilon_0\hbar^2}{m_e\, e^2} \tag{225}$$

The velocity $\vec{v}_1$ of the displacement current $\vec{I}_{el1}$ is $1.216\ 10^{-28}\ m/s$ and is the same for all masses. All atoms in the objects expand with the universe.

The magnetic flux density $\vec{B}_2$ in Equation (217) is caused by the mass m_2. It results from the magnetic field strength. For a point-like magnetic charge in space, the magnetic field strength is equivalent to the well-known electric field strength at a distance r:

$$\vec{H}_2 = \frac{\vec{Q}_{mg\,2}}{4\pi\mu_0\, r^2} \tag{226}$$

In the Earth-Moon system and in many other gravitational systems, the masses can be regarded as point-like with respect to their gravitational effect due to their great distance.

This yields the magnetic flux density $\vec{B}_2$ with the absolute permeability μ and the distance r between the Earth and the Moon:

$$\vec{B}_2 = \mu\vec{H}_2 = \mu_0\mu_r\vec{H}_2 = \mu_0\mu_r\frac{N_{p2}p}{4\pi\mu_0\, r^2} = \mu_r\frac{N_{p2}p}{4\pi\, r^2} \tag{227}$$

The number of magnetic charges N_{p2} is in turn calculated using the object mass m_2 as well as the proton mass m_p and the electron mass m_e. It corresponds to the number of electric charges on this mass:

$$N_{p2} = \frac{m_2}{m_p+m_e} \tag{228}$$

For the mass m_2 of the moon, this yields $N_{p2} = 4.393\ 10^{49}$ magnetic elementary charges. Equation (228) is supported by the structure of the proton as described in Chapter 17.

The material-dependent permeability μ_r is that of a vacuum and, by definition, has the value 1:

$$\mu_r = 1 \tag{229}$$

The magnetic flux density $\vec{B}_2$ according to Equation (227) is then $\vec{B}_2 = 1.428\ 10^{15}\ Vs/m^2$.

The resulting force according to equation (209) due to the electric displacement current $\vec{I}_{el\,1}$ in the magnetic field with flux density $\vec{B}_2$ can then be written as follows:

$$\vec{F}_{res1} = \vec{Q}_{el\,1}\,\vec{v}_1\,\vec{B}_2 = N_{e1}e\frac{r_{bohr}}{t_{uni}}\frac{N_{p2}}{4\pi\,r^2}p = \frac{m_1}{m_p+m_e}e\frac{r_{bohr}}{t_{uni}}\frac{\frac{m_2}{m_p+m_e}}{4\pi\,r^2}p \tag{230}$$

$$\vec{F}_{res1} = \frac{m_1 m_2}{\left(m_p+m_e\right)^2}\frac{r_{bohr}}{t_{uni}}\frac{ep}{4\pi\,r^2} \approx \frac{m_1 m_2}{m_p{}^2}\frac{r_{bohr}}{t_{uni}}\frac{ep}{4\pi\,r^2} \tag{231}$$

$$\vec{F}_{res1} = \vec{Q}_{el\,1}\,\vec{v}_1\,\vec{B}_2 = 5.722\ 10^{32} \times\ 1.216\ 10^{-28}\ \times 1.428\ 10^{15}\ N$$

$$\boldsymbol{\vec{F}_{res1} = 9.936\ 10^{19} N}$$

Using Equation (218), the resulting force $\vec{F}_{res2}$ exerted by the electric displacement current $\vec{I}_{el\,2}$ in the magnetic field with flux density $\vec{B}_1$ is to be determined (Fig. 10):

The total electric charge $\vec{Q}_{el\,2}$ on the mass m_2 is given by the number of individual charges. The number of individual charges, in turn, is calculated using the object mass m_2 as well as the proton mass m_p and the electron mass m_e:

$$\vec{Q}_{el\,2} = N_{e2}e \qquad \text{where} \qquad N_{e2} = \frac{m_2}{m_p+m_e} \tag{232}$$

For the Moon with mass m_2, this results in $N_{e2} = 4.393\ 10^{49}$ elementary electric charges. The total electric charge according to (232) is accordingly $\vec{Q}_{el\,2} = 7.039\ 10^{30}\ As$.

The velocity $\vec{v}_2$ of the displacement current $\vec{I}_{el2}$ resulting from the atomic expansion with the expansion of the universe is extremely slow and corresponds to the ratio of the Bohr radius r_{bohr} to the age of the universe t_{uni}.

$$\vec{v}_2 = \frac{r_{bohr}}{t_{uni}} \tag{233}$$

The velocity $\vec{v}_2$ of the displacement current $\vec{I}_{el2}$ is $1.216\ 10^{-28}\ m/s$ and is the same for all masses. All atoms in the objects expand with the universe.

The magnetic flux density $\vec{B}_1$ in Equation (218) is generated by the mass m_1. It is determined by the magnetic field strength. For a point magnetic charge in space, the magnetic field strength is equivalent to the well-known electric field strength at a distance r:

$$\vec{H}_1 = \frac{\vec{Q}_{mg\,1}}{4\pi\mu_0\, r^2} \tag{234}$$

In the Earth-Moon gravitational system, the masses can be considered point-like with respect to their gravitational effect due to their great distance.

This yields the magnetic flux density $\vec{B}_1$ with the absolute permeability μ and the distance$r$ between the Earth and the Moon:

$$\vec{B}_1 = \mu\vec{H}_1 = \mu_0\mu_r\vec{H}_1 = \mu_0\mu_r\frac{N_{p1}}{4\pi\mu_0\, r^2}p = \mu_r\frac{N_{p1}}{4\pi\, r^2}p \tag{235}$$

The number of magnetic charges N_{p1} is in turn calculated using the object mass m_1 as well as the proton mass m_p and the electron mass m_e. It corresponds to the number of electric charges on this mass:

$$N_{p1} = \frac{m_1}{m_p+m_e} \tag{236}$$

For the mass m_1 of the Earth, this results in $N_{p1} = 3.571\ 10^{51}$ elementary magnetic charges.

The material-dependent permeability μ_r is that of a vacuum and, by definition, has the value 1.

$$\mu_r = 1 \tag{237}$$

The magnetic flux density $\vec{B}_1$ according to Equation (235) is then $\vec{B}_1 = 1.161\ 10^{17}\ Vs/m^2$.

The resulting force $\vec{F}_{res2}$ according to Equation (218) due to the electric displacement current $\vec{I}_{el\,2}$ in the magnetic field with flux density $\vec{B}_1$ can then be written as follows:

$$\vec{F}_{res2} = \vec{Q}_{el\,2}\,\vec{v}_2\,\vec{B}_1 = N_{e2}e\frac{r_{bohr}}{t_{uni}}\frac{N_{p1}}{4\pi\, r^2}p = \frac{m_2}{m_p+m_e}e\frac{r_{bohr}}{t_{uni}}\frac{\frac{m_1}{m_p+m_e}}{4\pi\, r^2}p \tag{238}$$

$$\vec{F}_{res2} = \frac{m_1 m_2}{(m_p+m_e)^2}\frac{r_{bohr}}{t_{uni}}\frac{ep}{4\pi\, r^2} \tag{239}$$

$$\vec{F}_{res2} = \vec{Q}_{el\,2}\, \vec{v}_2\, \vec{B}_1 = 7.039\ 10^{30} \times\ 1.216\ 10^{-28}\ \times 1.161\ 10^{17}\ N$$

$$\vec{\boldsymbol{F}}_{\boldsymbol{res2}} = \mathbf{9.936\ 10^{19} N}$$

The gravitational force during the expansion of the universe is simply the sum of the first resultant force $\vec{F}_{res1}$ and the second force of equal magnitude $\vec{F}_{res2}$:

$$\vec{\boldsymbol{F}}_{\boldsymbol{grav}} = \vec{F}_{res1} + \vec{F}_{res2} = 9.936\ 10^{19}N + 9.936\ 10^{19}N = \mathbf{1.987\ 10^{20}\ N} \quad (239a)$$

The result from Equation (239a) can easily be verified using the gravitational force of Newton, which acts between the Earth and the Moon.

The following calculations using equations (240) through (255) relate to the theoretical possibility that the universe pulsates and that magnetic monopole charges exist in space during the contraction of the universe (see Fig. 11).

Equation (222) is used to determine the resulting force $\vec{F}_{res1}$ due to the magnetic displacement current $\vec{I}_{mg\,1}$ in the electric field with flux density $\vec{D}_2$ (Figure 11). The total magnetic charge $\vec{Q}_{mg\,1}$ on the mass m_1 is given by the number of individual charges. The number of individual charges, in turn, is calculated using the object mass m_1 as well as the proton mass m_p and the electron mass m_e:

$$\vec{Q}_{mg\,1} = N_{p1} p \qquad \text{where} \qquad N_{p1} = \frac{m_1}{m_p + m_e} \quad (240)$$

For the Earth with mass m_1, this yields $N_{p1} = 3.571\ 10^{51}$ magnetic elementary charges. The total magnetic charge according to (240) is accordingly $\vec{Q}_{mg\,1} = 2.156\ 10^{35}\ Vs$.

The velocity $\vec{v}_1$ of the displacement current $\vec{I}_{mg1}$ resulting from atomic miniaturization with the contraction of the universe is extremely slow and corresponds to the ratio of the Bohr radius r_{bohr} to the age of the universe t_{uni}.

$$\vec{v}_1 = \frac{r_{bohr}}{t_{uni}} \quad (241)$$

The velocity $\vec{v}_1$ of the displacement current $\vec{I}_{mg1}$ is $1.216\ 10^{-28}\ m/s$ and is the same for all masses. All atoms in the objects contract with the universe.

The electric flux density $\vec{D}_2$ in Equation (222) is caused by the mass m_2. It is derived from the electric field strength.

For a point-like electric charge in space, the following applies to the electric field strength at a distance r:

$$\vec{E}_2 = \frac{\vec{Q}_{el\,2}}{4\pi\varepsilon_0\, r^2} \tag{242}$$

In the Earth-Moon gravitational system, the masses can be considered point-like with respect to their gravitational effect due to their great distance.

This yields the electric flux density $\vec{D}_2$ with the absolute permittivity ε and the distance r between the Earth and the Moon:

$$\vec{D}_2 = \varepsilon\vec{E}_2 = \varepsilon_0\varepsilon_r\vec{E}_2 = \varepsilon_0\varepsilon_r\frac{N_{e2}e}{4\pi\varepsilon_0\, r^2} = \varepsilon_r\frac{N_{e2}e}{4\pi\, r^2} \tag{243}$$

The number of electric charges N_{e2} is in turn calculated using the object's mass m_2, the proton mass m_p, and the electron mass m_e. It corresponds to the number of magnetic charges on this mass:

$$N_{e2} = \frac{m_2}{m_p+m_e} \tag{244}$$

For the mass m_2 of the moon $N_{e2} = 4.393\;10^{49}$, this results in elementary electric charges.

The material-dependent permittivity ε_r, just like the material-dependent permeability μ_r, corresponds exactly to that of a vacuum and, by definition, has the value 1:

$$\varepsilon_r = 1 \tag{245}$$

The electric flux density $\vec{D}_2$ according to Equation (243) is then $\vec{D}_2 = 3.791\;10^{12}\;As/m^2$.

The resulting force according to equation (214) due to the magnetic displacement current $\vec{I}_{mg\,1}$ in the electric field with flux density $\vec{D}_2$ can then be written as follows:

$$\vec{F}_{res1} = \vec{Q}_{mg\,1}\,\vec{v}_1\,\vec{D}_2 = N_{p1}p\frac{r_{bohr}}{t_{uni}}\frac{N_{e2}}{4\pi\, r^2}e = \frac{m_1}{m_p+m_e}p\frac{r_{bohr}}{t_{uni}}\frac{\frac{m_2}{m_p+m_e}}{4\pi\, r^2}e \tag{246}$$

$$\vec{F}_{res1} = \frac{m_1 m_2}{(m_p+m_e)^2}\frac{r_{bohr}}{t_{uni}}\frac{ep}{4\pi\, r^2} \tag{247}$$

$$\vec{F}_{res1} = \vec{Q}_{mg\,1}\,\vec{v}_1\,\vec{D}_2 = 2.156\;10^{35} \times 1.216\;10^{-28} \times 3.791\;10^{12}\;N$$

$$\boldsymbol{\vec{F}_{res1} = 9.936\;10^{19} N}$$

Using Equation (223), the resulting force $\vec{F}_{res2}$ due to the magnetic displacement current $\vec{I}_{mg\,2}$ in the electric field with flux density $\vec{D}_1$ is to be determined (Fig. 11):

The total magnetic charge $\vec{Q}_{mg\,2}$ on the mass m_2 is given by the number of individual charges. The number of individual charges, in turn, is calculated using the object mass m_2 as well as the proton mass m_e and the electron mass m_p:

$$\vec{Q}_{mg\,2} = N_{p2}p \qquad \text{where} \qquad N_{p2} = \frac{m_2}{m_p+m_e} \tag{248}$$

For the Moon with mass m_2, this yields $N_{p2} = 4.393\ 10^{49}$ magnetic elementary charges. The total magnetic charge according to (248) is then $\vec{Q}_{mg\,2} = 2.652\ 10^{33}\ Vs$.

The velocity $\vec{v}_2$ of the displacement current $\vec{I}_{mg2}$ resulting from atomic miniaturization due to the contraction of the universe is extremely slow and corresponds to the ratio of the Bohr radius r_{bohr} to the age of the universe t_{uni}.

$$\vec{v}_2 = \frac{r_{bohr}}{t_{uni}} \tag{249}$$

The velocity $\vec{v}_2$ of the displacement current $\vec{I}_{mg2}$ is $1.216\ 10^{-28}\ m/s$ and is the same for all masses. All atoms in the objects contract with the universe.

The electric flux density $\vec{D}_1$ in Equation (223) is caused by the mass m_1. It is derived from the electric field strength. For a point-like electric charge in space, the electric field strength at a distance$r$ is given by:

$$\vec{E}_1 = \frac{\vec{Q}_{el\,1}}{4\pi\varepsilon_0\, r^2} \tag{250}$$

In the Earth-Moon gravitational system, the masses can be considered point-like with respect to their gravitational effect due to their great distance.

This yields the electric flux density $\vec{D}_1$ with the absolute permittivity ε and the distance r between the Earth and the Moon:

$$\vec{D}_1 = \varepsilon\vec{E}_1 = \varepsilon_0\varepsilon_r\vec{E}_1 = \varepsilon_0\varepsilon_r\frac{N_{e1}}{4\pi\varepsilon_0\, r^2}e = \varepsilon_r\frac{N_{e1}}{4\pi\, r^2}e \tag{251}$$

The number of electric charges N_{e1} is in turn calculated using the object mass m_1, the proton mass m_p, and the electron mass m_e. It corresponds to the number of magnetic charges on this mass:

$$N_{e1} = \frac{m_1}{m_p+m_e} \quad (252)$$

For the mass m_1 of the Earth, this results in $N_{e1} = 3.571\ 10^{51}$ elementary magnetic charges.

The material-dependent permittivity ε_r corresponds to that of a vacuum and, by definition, has the value 1.

$$\varepsilon_r = 1 \quad (253)$$

The electric flux density $\vec{D}_1$ according to Equation (243) is then $\vec{D}_1 = 3.082\ 10^{14}\ As/m^2$.

The resulting force $\vec{F}_{res2}$ according to Equation (223) due to the electric displacement current $\vec{I}_{el\,2}$ in the magnetic field with flux density $\vec{B}_1$ can then be written as follows:

$$\vec{F}_{res2} = \vec{Q}_{mg\,2}\,\vec{v}_2\,\vec{D}_1 = N_{e2}p\frac{r_{bohr}}{t_{uni}}\frac{N_{e1}}{4\pi\,r^2}e = \frac{m_2}{m_p+m_e}p\frac{r_{bohr}}{t_{uni}}\frac{\frac{m_1}{m_p+m_e}}{4\pi\,r^2}e \quad (254)$$

$$\vec{F}_{res2} = \frac{m_1 m_2}{(m_p+m_e)^2}\frac{r_{bohr}}{t_{uni}}\frac{ep}{4\pi\,r^2} \quad (255)$$

$$\vec{F}_{res2} = \vec{Q}_{mg\,2}\,\vec{v}_2\,\vec{D}_1 = 2.652\ 10^{33} \times 1.216\ 10^{-28}\ x\ 3.082\ 10^{14}\ N$$

$$\boldsymbol{\vec{F}_{res2} = 9.936\ 10^{19} N}$$

Even during the contraction of the universe, the gravitational force is simply the sum of the first resulting force $\vec{F}_{res1}$ and the second force of equal magnitude $\vec{F}_{res2}$:

$$\boldsymbol{\vec{F}_{grav}} = \vec{F}_{res1} + \vec{F}_{res2} = 9.936\ 10^{19}N + 9.936\ 10^{19}N = \boldsymbol{1.987\ 10^{20}\ N} \quad (255a)$$

The result of equation (255a) is consistent with that of equation (239a) and can again be verified using the gravitational force of Newton acting between the Earth and the Moon.

The gravitational force during the theoretically considered contraction of the universe is exactly as large as during expansion, and it remains attractive.

Equations (231), (239), (247), and (255) contain elementary magnetic charges that are indispensable for the electromagnetic description of gravity. The obvious conclusion is: **They exist within the atoms.**

For the Earth-Moon gravitational system with a center-to-center distance of$r$ between the two objects, the calculated gravitational force is $1.987\ 10^{20} N$. It is left to the reader to determine this gravitational force according to Newton. The usual negative sign of the gravitational force follows from the definition that repulsive forces are positive and is not taken into account here.

The slowed flow of time in the gravitational field is an indication that the induced force inhibits its cause, namely the displacement current.

The concept of spacetime curvature as the cause of gravity should be rejected. Newton postulated the gravitational force, and Einstein gave this force a geometric cause. This is merely a shifting of the problem; neither of them explains the true cause. The general theory of relativity is based on the equality of inertial and gravitational mass. From the equation $F = m_{inertial}\ a = G\ M\ m_{massiv}/r^2$, it immediately follows that $a = G\ M\ /r^2$. The inertial mass cancels out against the gravitational mass. All masses should experience the same acceleration from rest in a homogeneous gravitational field of mass M, and thereby move along the same trajectory and collide at the same final velocity. However, there are no homogeneous gravitational fields, as these are bound to object masses with typical geometries. Furthermore, the reaction of the falling masses on the object attracting them must be taken into account. It follows that heavy masses have a lower final velocity upon impact than lighter ones: thus, not all masses fall at the same speed. I have pointed this out many times.[45] At best, massless test objects without reaction could exhibit such behavior. However, these do not exist in nature. The measurement results showing that, for example, a down feather falls as fast as a steel ball in a vacuum are simply not precise enough to detect the difference. This is because the falling masses in the experiments are always extremely small compared to the mass attracting them. As explained in Chapter 13, the Newtonian gravitational constant G, from which the Einstein constant κ of GR follows, cannot be constant either.

Sometimes, well-funded organizations and corporate interests stand in the way of the truth. If one wanted to detect gravitational waves, it would be possible to do so even if measurement errors or other disturbances occurred in the extremely sensitive sensors. This is similar to the situation where pharmaceutical companies exploit a dangerous virus to sell vaccines. If a manipulated virus isn't dangerous enough, one could engineer the danger. Practice shows that the media can also unintentionally contribute to spreading misinformation. This can create a completely false picture in the public eye.

---

[45] Andreas Gimsa, The Beauty of Nature, 2nd edition, Gieselmann Druck- und Medienhaus, Potsdam, 2014, ISBN 978-3-923830-94-7, p. 11

19. Gravity as a One-Dimensional Radiation Phenomenon [46]

In the following derivation, time is derived from the thermal radiation of one-dimensional emitters using quantum physics. It can thus be described as a one-dimensional radiation phenomenon. One-dimensional emitters in which standing waves propagate could exist in the universe.

The radiation mechanism is to be captured mathematically. To this end, the energy density is calculated for the one-dimensional radiator.[47] The energy within the radiator is stored in electromagnetic natural oscillations, the modes. The probability of a mode being occupied by bosons $f(\nu,\mathcal{T})$ is given by the Bose-Einstein distribution function:

$$f(\nu,\mathcal{T}) = \frac{1}{e^{h\nu/k\mathcal{T}}-1} \tag{256}$$

For the quantum mechanical energy per mode ϵ_{qm} , equation (257) applies, with the energy$h\nu$ of a boson with integer spin (photon or ether photon) in thermal equilibrium. Here, h is Planck's constant, ν is the frequency, k is Boltzmann's constant, and $\mathcal{T}$ is the temperature of the 1D radiation:

$$\epsilon_{qm} = h\nu\, f(\nu,\mathcal{T}) = \frac{h\nu}{e^{h\nu/k\mathcal{T}}-1} \tag{257}$$

Using the quantum mechanical energy ϵ_{qm}, the length L of the radiator, and the spectral mode density $g(\nu) = \frac{2L}{c}$, Equation (258) yields the spectral energy density $u(\nu,\mathcal{T})$ of the one-dimensional radiation law.

The spectral mode density $g(\nu)$ includes the factor 2 because there are two directions of propagation for the wave.

$$u(\nu,\mathcal{T}) = \frac{g(\nu)\,\epsilon_{qm}}{2\,L} = \frac{1}{c}\frac{h\nu}{e^{h\nu/k\mathcal{T}}-1} \tag{258}$$

The total energy density $u(\mathcal{T})$ of the one-dimensional blackbody is obtained by integrating the spectral energy density $u(\nu,\mathcal{T})$ over the entire frequency range:

$$u(\mathcal{T}) = \int_0^\infty u(\nu,\mathcal{T})\, d\nu = \frac{1}{c}\int_0^\infty \frac{h\nu}{e^{h\nu/k\mathcal{T}}-1} d\nu \tag{259}$$

Using the substitutions $x = \frac{h\nu}{k\mathcal{T}}$ and $d\nu = \frac{k\mathcal{T}}{h} dx$, the energy density $u(\mathcal{T})$ follows:

$$u(\mathcal{T}) = \frac{1}{c}\int_0^\infty \frac{x\, k\mathcal{T}}{e^x-1}\frac{k\mathcal{T}}{h} dx = \frac{1}{c}\frac{k^2\mathcal{T}^2}{h}\int_0^\infty \frac{x}{e^x-1} dx = \frac{\pi^2}{6}\frac{k^2\mathcal{T}^2}{c\,h} = \frac{\pi}{12}\frac{k^2\mathcal{T}^2}{c\,\hbar} \tag{260}$$

For the total internal energy $U(\mathcal{T})$ of the radiator, applies over its length L:

---

[46] Andreas Gimsa, Gravity: Part of the Beauty of Nature, Part 2, Kuss GmbH, Potsdam, 2016, ISBN 978-3-00-052808-8

[47] Wallraff, Lecture Notes Spring Semester 2011, Physics IV, Introduction to Quantum Mechanics, Chapter V, Thermal Radiation, One-Dimensional Radiation Case, ETH Zurich, pp. 51–57

$$U(\mathcal{T}) = u(\mathcal{T})L = \frac{\pi}{12}\frac{k^2\mathcal{T}^2}{c\hbar}L = \frac{\pi}{12}\frac{k^2\mathcal{T}^2}{c\hbar}cT = \frac{\pi^2}{6}\frac{k^2\mathcal{T}^2}{\hbar}t_b \tag{261}$$

Here, $\boldsymbol{L = c\,T_b = 2\pi\,c\,t_b}$ (see Fig. 12). t_b is to be the time interval during the motion of the object that emits the one-dimensional radiation (but in the present case absorbs the gravitational field). From Wick rotation, the expression $e^{-\frac{E\,it}{\hbar}}$ is known. Here, an ensemble of oscillators consists of a superposition of ground states and evolves over time with the Hamilton operator.[48] The relative probability of encountering an oscillator with a specific energy corresponds to the phase change of a basis state with that energy. Each basis state of an oscillator corresponds to a discrete energy value with a specific frequency. The following applies:

$$e^{-\frac{E}{k\mathcal{T}}} = e^{-\frac{E\,it}{\hbar}} \tag{262}$$

To eliminate unwanted oscillations, the substitution $\tau = it$ is used, which calculates with the modulus τ of the imaginary time:

$$\frac{E}{k\,\mathcal{T}} = \frac{E\,\tau}{\hbar} \qquad\qquad \mathcal{T} = \frac{\hbar}{k\,\tau} \tag{263}$$

Here, $\mathcal{T}$ is the temperature of one-dimensional radiation. The square of the imaginary time is replaced by the square of the time interval τ:

$$k^2\mathcal{T}^2 = \frac{\hbar^2}{\tau^2} \tag{264}$$

For $u(\mathcal{T})$, the following simple relationship holds in connection with the one-dimensional radiation constant α_{1d}:

$$u(\mathcal{T}) = \alpha_{1d}\,\mathcal{T}^2 \tag{265}$$

Using equations (260) and (265), α_{1d} can be determined immediately:

$$\alpha_{1d} = \frac{\pi}{12}\frac{k^2}{c\,\hbar} \tag{266}$$

The following applies to the total radiant power P_{1d} of the one-dimensional radiator:

$$P_{1d} = 2\pi\,c\,u(\mathcal{T}) = 2\pi\,c\,\frac{\pi}{12}\frac{k^2\mathcal{T}^2}{c\,\hbar} = \frac{\pi^2}{6}\frac{k^2\mathcal{T}^2}{\hbar} = \sigma_{1d}\mathcal{T}^2 \tag{267}$$

Consequently, the following applies to the one-dimensional radiation constant σ_{1d}:

$$\sigma_{1d} = \frac{\pi^2}{6}\frac{k^2}{\hbar} \tag{268}$$

[48] Pelster, Lecture Notes Spring Semester 2004, Bose-Einstein Condensation, Chapter 2, Quantum Statistics of a Particle, 2.3 Wick Rotation, University of Duisburg-Essen, p. 18

The infinite series $\sum_{n=1}^{\infty}\frac{1}{n^2}$ converges and, as Leonhard Euler proved, can be written as $\pi^2/6$.

This expression can arise, for example, as a factor in the ground state energy when all the energies of an electron in a hydrogen atom are summed across all possible energy levels. There are infinitely many of these. For example, the concise expression $\sum_{n=1}^{\infty}\frac{1}{n^2}$ also appears in the Casimir effect.

$$P_{1d} = \frac{\pi^2}{6}\frac{k^2\mathcal{T}^2}{\hbar} = \sum_{n=1}^{\infty}\frac{1}{n^2}\frac{k^2\mathcal{T}^2}{\hbar} \tag{269}$$

The expression from Equation (264) is now substituted into Equation (269):

$$P_{1d} = \sum_{n=1}^{\infty}\frac{1}{n^2}\frac{\hbar}{\tau^2} \tag{270}$$

It is assumed that the one-dimensional radiators in the object masses can absorb (or emit) the time radiation from their internal energy only in the ground state $(n = 1)$. Then, using Equation (270):

$$P_{1d\,n=1} = \frac{k^2\mathcal{T}^2}{\hbar} = \frac{k^2\mathcal{T}^2}{\hbar} = \frac{\hbar}{\tau^2} \tag{271}$$

For two masses interacting with emitters at $N_{1\hbar}$ and $N_{2\hbar}$, the following holds from (263):

$$P_{1d} = -(P_{1d1} + P_{1d2}) = -\hbar\left(\frac{N_{1\hbar}}{\tau^2} + \frac{N_{2\hbar}}{\tau^2}\right) \tag{272}$$

$N_{1\hbar}$ and $N_{2\hbar}$ is the effective number of one-dimensional radiators. For two objects in radiative equilibrium, it can be calculated as follows:

$$N_{1\hbar} = N_1\sqrt{\frac{\alpha_{grav}}{\alpha_{em}}} \qquad\qquad N_{2\hbar} = N_2\sqrt{\frac{\alpha_{grav}}{\alpha_{em}}} \tag{273}$$

N_1 and N_2 in (273) are the quantities of geometrically averaged charge masses of electrons and protons contained in the two masses:

$$N_1 = \frac{m_1}{\sqrt{m_e m_p}} \qquad\qquad N_2 = \frac{m_2}{\sqrt{m_e m_p}} \tag{274}$$

For the gravitational interaction constant α_{grav}, the following applies:

$$\alpha_{grav} = \frac{G\, m_e m_p}{\hbar\, c} \tag{275}$$

For the interaction constant of the electromagnetic force α_{em}, the well-known relationship holds with the elementary electric charge$e$ and, alternatively, with the elementary magnetic charge p:

$$\alpha_{em} = \frac{e^2}{4\pi\,\varepsilon_0\,c\,\hbar} = \frac{p^2}{4\pi\,\mu_0\,c\,\hbar} \tag{276}$$

Equations (274), (275), and (276) are substituted into Equation (273):

$$N_{1\hbar} = N_1\sqrt{\frac{\alpha_{grav}}{\alpha_{em}}} = \frac{m_1}{\sqrt{m_e m_p}}\sqrt{\frac{\frac{G\,m_e m_p}{\hbar\,c}}{\frac{e^2}{4\pi\,\varepsilon_0 c\,\hbar}}} = \frac{m_1}{e}\sqrt{4\pi\,\varepsilon_0\,G} = \frac{m_1}{p}\sqrt{4\pi\,\mu_0\,G} \tag{277}$$

$$N_{2\hbar} = N_2\sqrt{\frac{\alpha_{grav}}{\alpha_{em}}} = \frac{m_2}{e}\sqrt{4\pi\,\varepsilon_0\,G} = \frac{m_2}{p}\sqrt{4\pi\,\mu_0\,G} \tag{278}$$

The following equality between the forces at a distance r_b between two electric or two magnetic elementary charges and the gravitational force holds in connection with the interaction constants α_{em} and α_{grav}. This astonishing relationship has been established, verified, and is exactly correct:

$$\alpha_{grav}\,\frac{e^2}{4\pi\varepsilon_0 {r_b}^2} = \alpha_{grav}\,\frac{p^2}{4\pi\mu_0 {r_b}^2} = \alpha_{em}\,G\,\frac{m_e m_p}{{r_b}^2} \tag{279}$$

A similar balance of forces applies between two masses, each possessing elementary charges of N:

$$\alpha_{grav}\,\frac{N_1 N_2\,e^2}{4\pi\varepsilon_0 {r_b}^2} = \alpha_{grav}\,\frac{N_1 N_2\,p^2}{4\pi\mu_0 {r_b}^2} = \alpha_{em}\,G\,\frac{m_1 m_2}{{r_b}^2} \tag{280}$$

Through multiplication, division, mathematical manipulation of the equations, several concise equations can be derived:

$$N_1 N_2 = \frac{m_1 m_2}{m_e m_p} \qquad N_{1\hbar} N_{2\hbar} = \frac{\alpha_{grav}}{\alpha_{em}}\frac{m_1 m_2}{m_e m_p} \qquad \frac{{N_{1\hbar}}^2 N_2}{{N_{2\hbar}}^2 N_1} = \frac{m_1}{m_2} \tag{281}$$

To determine the gravitational energy of a fall from rest until impact, the gravitational power given by Equation (272) must be integrated over the time interval of the fall t_{fall} and the shortest time interval t_b at impact:

$$E_{1\mathrm{d}} = E_{\mathrm{grav}} = -\int_{t_{1b}}^{t_{fall}} P_{1d1}\,d\tau - \int_{t_{2b}}^{t_{fall}} P_{1d2}\,d\tau \tag{282}$$

Thus, the total gravitational energy of both masses is given by:

$$E_{1\mathrm{d}} = E_{\mathrm{grav}} = \int_{t_{1b}}^{t_{fall}} \frac{N_{1\hbar}\hbar}{\tau^2}\,d\tau + \int_{t_{2b}}^{t_{fall}} \frac{N_{2\hbar}\hbar}{\tau^2}\,d\tau \tag{283}$$

$$E_{1\mathrm{d}} = E_{\mathrm{grav}} = \frac{N_{1\hbar}\hbar}{\tau}\Big|_{t_{1b}}^{t_{fall}} + \frac{N_{2\hbar}\hbar}{\tau}\Big|_{t_{2b}}^{t_{fall}} = \frac{N_{1\hbar}\hbar}{t_{fall}} - \frac{N_{1\hbar}\hbar}{t_{1b}} + \frac{N_{2\hbar}\hbar}{t_{fall}} - \frac{N_{2\hbar}\hbar}{t_{2b}} \tag{284}$$

In Equation (284), the terms involving the fall time t_{fall} drop out, since they are very small compared to the terms involving the time interval t_b during motion:

$$E_{1\mathrm{d}} = E_{\mathrm{grav}} = -\hbar\left(\frac{N_{1\hbar}}{t_{1b}} + \frac{N_{2\hbar}}{t_{2b}}\right) \tag{285}$$

As expected, the gravitational energy E_{grav} has a negative sign. Equation (285) has been verified with many examples involving two-mass systems. It yields correct results.

How can the time intervals t_b of the objects in motion be calculated? This is already shown by equations (100) through (103):

$$t_{1b} = -\frac{\Delta t_1\, m_1}{\Delta m_1} \qquad t_{2b} = -\frac{\Delta t_2\, m_2}{\Delta m_2} \tag{286}$$

The change in the time intervals t_b of the moving object state compared to the time intervals t_0 in the rest state can be written as follows:

$$\Delta t_1 = -\frac{N_{1\hbar}\,\hbar}{m_1\, c^2} = t_{1b} - t_{10} \qquad \Delta t_2 = -\frac{N_{2\hbar}\,\hbar}{m_2\, c^2} = t_{2b} - t_{20} \tag{287}$$

The changes in the time intervals Δt for both masses are identical and can be written using transformations from equations (273) to (278) with the Planck time t_{Planck} and the fine-structure constant α:

$$\Delta t = \Delta t_1 = \Delta t_2 = \frac{t_{Planck}}{\sqrt{\alpha}} \qquad \text{where } t_{planck} = \sqrt{\frac{\hbar\, G}{c^5}} \tag{288}$$

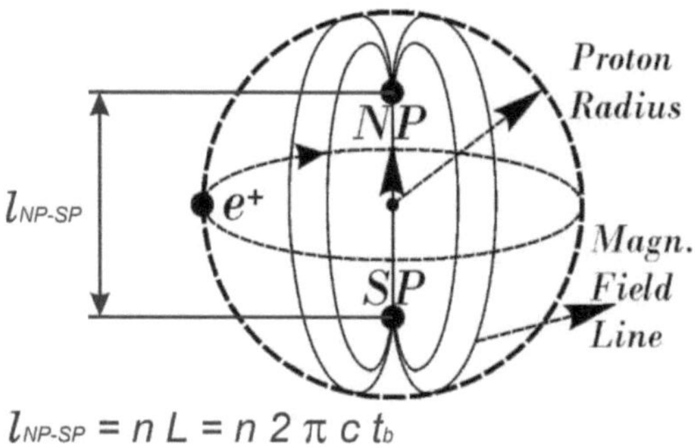

Figure 12: *Model of the proton according to Fig. 9 with variable distance* s_{NP-SP}

The distance l_{NP-SP} in Fig. 12 is not a fixed distance between the north and south poles of the proton, but rather, according to the principles of quantum physics, a probable distance on the order of the proton radius. In principle, it could be composed of$n$ radiators of length L. The one-dimensional radiators could exist on the central field line, Fig. 12. The extremely short time interval t_b is still many orders of magnitude smaller than the light travel time for a distance corresponding to the proton radius. Any time interval t that is reduced on the moving object due to gravity is reduced proportionally to the time interval t_b. Consequently, time is eliminated in the presence of gravity. Similarly, when two masses are separated—as occurs, for example, when lifting a stone off the ground—time must be created.

In this process, the protons would have to become larger. This can also be observed in the expansion of the universe, in which massive galaxies move away from one another: new time is constantly being created in the expanding universe.

Equations (286) and (287) yield the relationships that allow the gravitational energy from Equation (285) to be expressed in terms of changes in object mass:

$$t_{1b} = \frac{N_{1\hbar}\,\hbar}{m_1\,c^2}\frac{m_1}{\Delta m_1} = \frac{N_{1\hbar}\,\hbar}{\Delta m_1\,c^2} \qquad t_{2b} = \frac{N_{2\hbar}\,\hbar}{m_2\,c^2}\frac{m_2}{\Delta m_2} = \frac{N_{2\hbar}\,\hbar}{\Delta m_2\,c^2} \tag{289}$$

$$E_{\text{grav}} = -(\Delta m_1 + \Delta m_2)\,c^2 \tag{290}$$

Equation (291) shows equation (285) in frequency notation for one-dimensional gravitational radiation:

$$E_{1d} = -h(N_{1\hbar}f_{1b} + N_{2\hbar}f_{2b}) \qquad \text{with} \quad f_{1b} = \frac{1}{2\pi t_{1b}} \quad f_{2b} = \frac{1}{2\pi t_{2b}} \tag{291}$$

Time radiation could be transmitted in the form of longitudinal waves, with only one dimension available for its propagation. Because only time information and no energy needs to be transmitted, normal photons are not required; instead, "only" virtual photons are needed. These could be the imaginary photons from Chapter 5, which are not spatial but temporal in nature.

Although I am very taken with this second quantum-physical derivation of gravity as a one-dimensional radiation phenomenon, a sound justification for the length L of the one-dimensional radiators (see Fig. 12) still needs to be provided. A direct measurement will not be possible; perhaps a clever reader has a good idea on this.

How does this fit together: gravity as an electromagnetic phenomenon arising from displacement currents as atoms expand during the expansion of the universe, and gravity as a one-dimensional radiation phenomenon of time?

It fits together just as well as the fact of quantized and equivalent energy emission in the form of photon emission of certain frequencies when electrons in the atom reach a lower energy level. The increasing kinetic energy of the objects due to gravity leads to a time cancellation; one-dimensional time radiation with the same energy is absorbed. The potential energy of the gravitational field is thereby reduced. As demonstrated in Chapter 12, the product of mass and time interval remains constant for every object involved in the gravitational process: $m_x t_x = const$. This also aligns with the energy-time uncertainty principle of quantum-physics.

20. Symmetries, Fundamental Forces, and their Unification [49]

Symmetries play an essential role in nature. Symmetrical structures are generally perceived as beautiful. Mathematicians and physicists even regard symmetries in the equations describing the world mathematically and physically as an indication of their correctness. As the British mathematician Godfrey Harold Hardy writes, "The mathematician's patterns, like the painter's or the poet's, must be beautiful; the ideas, like the colors or the words, must fit together in a harmonious way. Beauty is the first test: there is no permanent place in the world for ugly mathematics." [50] Emmy Noether provided a very interesting example of symmetries in physics.

She observed that certain system properties remain unchanged under transformations. From the invariance (unchangeability) of the laws of nature under transformations of time, space, and direction, Emmy Noether derived the laws of conservation of energy, momentum, and angular momentum. These symmetries and their conservation laws form the foundation of physics.

In this chapter, we will examine other important symmetries.

<u>Symmetry between the strong force and the gravitational force</u>

This newly formulated symmetry is remarkable because it connects the relationship between the strongest and the weakest forces with the geometry of "very large" and "very small." This will be discussed further in Chapter 21. Equation (292) shows us the connection between the strong force and the gravitational force in relation to the radius of the universe and the proton radius:

$$\frac{r_{uni}}{r_p} = \frac{\alpha_{strong}}{\alpha_{grav}} = \frac{F_{strong}}{F_{grav}} \qquad (292)$$

From Equation (292) follows a law that I would like to call ***the lever law of the universe***:

$$\boldsymbol{F_{strong}\, r_p = F_{grav}\, r_{uni}} \quad E_1 = E_2 = 4.74\ 10^{-21}\, J = m_\nu\, c^2\ ? \qquad (292a)$$

The strong force between two charges in the proton, multiplied by the proton radius, is as large as the gravitational force between these charges multiplied by the radius of the universe. This also represents an energy balance that could point to the rest mass of the electron neutrino.

---

[49] Andreas Gimsa, The Unification of the Fundamental Interactions, International Journal of Scientific Research and Management, Volume 09, Issue 05, May 2021, DOI: 10.18535/ijsrm/v9i05.p01

[50] Hardy, Godfrey Harold, A Mathematician's Apology, University Press, Cambridge 1940, ISBN 978-0-521-42706-7

The strong interaction constant $\alpha_{strong} = 0.5$ represents a limit up to which perturbation theory is applicable. At a value of $\alpha_{strong} > 0.5$, individual quanta can no longer be identified.

Consequently, the proton exists as a whole, and it should therefore be possible to determine the radius of the entire proton (proton radius) using this limit.

$$\alpha_{strong} = \sqrt{\frac{\hbar}{m_p\, r_p\, c}} = 0.5 \qquad (293)$$

This yields the simple and elegant relationship for the proton radius:

$$r_p = \frac{4\hbar}{c\, m_p} \qquad (294)$$

The radius of the proton depends directly on its mass. Since both the speed of light and the reduced Planck constant are physical constants, this radius can only change with the proton mass.

As shown in Chapter 13, this mass decreases with the expansion of the universe. Consequently, the proton radius increases over time!

Symmetry of Electric and Magnetic Elementary Charges

There are not many sources that establish a formal relationship for the magnetic elementary charge$p$. For the force ratio between two magnetic and two electric elementary charges, the following relationship was given in "Gerthsen Physics": [51]

$$\frac{F_{mg}}{F_{el}} = \frac{p^2 \mu_0\, \varepsilon_0}{e^2} = \frac{1}{{\alpha_{em}}^2} \qquad (295)$$

This would mean that two magnetic elementary charges interact much more strongly with each other (factor of approx. 20,000) than two electric ones.

If, on the other hand, one divides a magnetic force with the magnetic elementary charge p by an electric force in a symmetric structure, one obtains the following formula instead of (295):

$$\frac{F_{mg}}{F_{el}} = \frac{\frac{p^2}{4\pi\mu_0 r^2}}{\frac{e^2}{4\pi\varepsilon_0 r^2}} = \frac{p^2 \varepsilon_0}{e^2 \mu_0} \qquad (296)$$

Since equations (295) and (296) contradict each other, they must be verified. To do this, the SI units are first analyzed.

[51] Dieter Meschede, Gerthsen Physics, 25th edition, Springer-Verlag Heidelberg, 2015, ISBN 978-3-662-45977-5, p. 981

Using the known formula $1/c = \sqrt{\varepsilon_0 \mu_0}$ and (295), the following would simply hold: $p = ec/\alpha_{em}$? Since e has the unit [As], the unit [Am] would have to be assigned to the monopole p.

However, according to Equation (296), using ε_0 in [As/Vm] and μ_0 in [Vs/Am] for p would yield the unit [Vs]. This seems plausible. Further analysis takes the electric and magnetic field strengths into account. For an electron experiencing a force $\vec{F}$, the electric field strength is: $\vec{E} = \vec{F}/e$. The electric field strength thus has the unit [V/m]. Now there should be an analogy such that a magnetic field strength $\vec{H} = \vec{F}/p$ exists at a monopole, which also experiences a force $\vec{F}$. The magnetic field strength has the unit [A/m]. However, this unit only holds if one substitutes the unit [Vs] from (296) for p. If one substitutes the unit [Am] from (295) for p, the unit of the magnetic field strength becomes [Vs/m²]. However, this is the unit of the magnetic flux density $\vec{B}$.

The aim is to investigate how large the force ratio F_{mg}/F_{el} actually is. To this end, the following experimental setup in a vacuum is proposed. It can only be considered from a theoretical standpoint, since magnetic monopoles, as presumed time generators, are not accessible to us in space. [52]

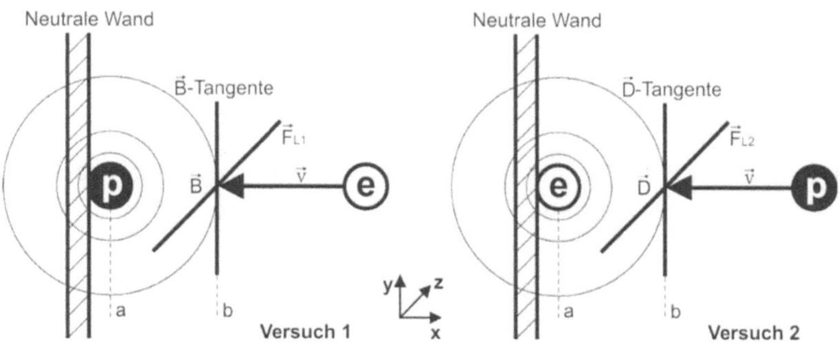

Figure 13: *Symmetrical effect of the Lorentz force*

The first experiment involves an elementary electric chargee e flying toward an elementary magnetic charge p —which is attached to an electrically and magnetically neutral wall—at a speed of $\vec{v}$ through the point in space on the b – line.

[52] Andreas Gimsa, The Impedance of Vacuum Part 1, Kuss GmbH Potsdam, 2015, ISBN 978-3-00-051924-6, p. 3

That is, for the electric charge, there is no external electric field $(\vec{E}=0)$, but only that of the magnetic elementary charge. The Lorentz force with a moving electric elementary charge is then at this point in space:

$$\vec{F}_{L1} = e\vec{E} + e(\vec{v}\times\vec{B}) = e(\vec{v}\times\vec{B}) \tag{297}$$

In the second experimental setup, the elementary electric charge e is now attached to the wall, and the elementary magnetic charge p flies at the same speed $\vec{v}$ through the point in space on the $b-$ line toward e. For the magnetic elementary charge, there is no external magnetic field $(\vec{H}=0)$, but only the electric field of the elementary charge e. Since this experimental setup is symmetric to the first one, the following formulation must apply here, with the moving magnetic elementary charge at the same spatial point:

$$\vec{F}_{L2} = p\vec{H} + p(\vec{v}\times\vec{D}) = p(\vec{v}\times\vec{D}) \tag{298}$$

Assuming point charges and a direction of motion perpendicular to the respective field lines, equations (297) and (298) simplify such that the vector products disappear and all vectors reduce to their magnitudes:

$$F_{L1} = e\,v\,B \tag{299}$$

$$F_{L2} = p\,v\,D \tag{300}$$

Since both experiments deal with exactly the same problem, the following holds:

$$F_{L1} = F_{L2} = e\,v\,B = e\,v\,\mu_0 H = p\,v\,D = p\,v\,\varepsilon_0 E \tag{301}$$

From this, the ratio of electric and magnetic field strengths can now be expressed as the wave impedance of the wave propagating in a vacuum:

$$\frac{E}{H} = \frac{e\,\mu_0}{p\,\varepsilon_0} = Z_0 = \sqrt{\frac{\mu_0}{\varepsilon_0}} \quad \text{or} \quad \frac{p\,\varepsilon_0}{e\,\mu_0} = \sqrt{\frac{\varepsilon_0}{\mu_0}} \tag{302}$$

Equation (302), in turn, can be converted into the ratio of magnetic force to electric force. The calculation formula is consistent with Equation (296).

$$\frac{F_{mg}}{F_{el}} = \frac{p^2\varepsilon_0}{e^2\mu_0} = 1 \tag{303}$$

The force between two magnetic monopoles is exactly as large as the force between two elementary electric charges. Equation (295) is incorrect when considered in SI units.

For the characteristic impedance of a vacuum, Equation (302) can be written very simply as:

$$Z_0 = \frac{p}{e} \tag{304}$$

Using the well-known equation (305), we can now easily derive a comparable formula for the magnetic elementary charge p :

$$e^2 = 4\pi \, \alpha_{em} \, c \, \varepsilon_0 \, \hbar = 2 \, \alpha \, c \, \varepsilon_0 \, h \tag{305}$$

$$p^2 = 4\pi \, \alpha_{em} \, c \, \mu_0 \, \hbar = 2 \, \alpha \, c \, \mu_0 \, h \tag{306}$$

Multiplying equation (305) by (306) yields the following concise expression after simplification:

$$h = \frac{1}{2 \, \alpha_{em}} ep \tag{307}$$

This leads, via (293), to the wonderful and descriptive equation (307a), which reveals the ratio of the Planck constant to the product of booth charges, ep:

$$\frac{h}{ep} = \frac{\alpha_{strong}}{\alpha_{em}} = \frac{F_{strong}}{F_{em}} \tag{307a}$$

Now, using equation (307), the Klitzing resistance $R_{Kl} = h/e^2$ can be expressed as follows:

$$R_{Kl} = \frac{1}{2 \, \alpha_{em}} \frac{p}{e} = \frac{\alpha_{strong}}{\alpha_{em}} \frac{p}{e} \tag{308}$$

It is well known that the energy of an oscillating system is quantized at specific energy levels at its frequency via Planck's constant h. Dirac predicted that the magnetic charge is quantized in integer multiples of $e/(2\alpha_{em})$ and that the force between two monopoles is $1/(4 \, {\alpha_{em}}^2) = 4{,}695$ times that between two electrons.[53] The energy quantization condition is applied to Equation (307): $Nep/(2\alpha_{em})$. The quantization condition for the Hall resistance via the Klitzing resistance according to Equation (308) is: $p/(N2\alpha_{em}e)$. Dividing the first quantization condition by the second immediately yields the necessary quantization of the electric charge: From N^2e^2, integer multiples of e follow for every electric charge: $Q_e = Ne$. Multiplying both quantum conditions yields the expression $p^2/(4{\alpha_{em}}^2)$. This means that the magnetic charge is quantized in integer multiples of $p/(2\alpha_{em})$: $Q_m = Np/(2\alpha_{em}) = \mu_r Np = 68.5 \, Np$.

We can already observe this quantisation in the strong force. It acts between an electric charge and a magnetic charge within the proton. As equation (204) also shows, the following holds: $F_{strong} = \mu_r ep/\left(4\pi\sqrt{\varepsilon_0\mu_0} \, r^2\right)$.

[53] P.A.M. Dirac, Quantized Singularities in the Electromagnetic Field, Proceedings of the Royal Society of London, Series A, Containing Papers of a Mathematical and Physical Character, 133 (821), DOI.ORG/10.1098/RSPA.1931.0130

The elementary magnetic charge p can now be conveniently calculated using equation (307):

$$p = 6.036\ 10^{-17}\ Vs \tag{309}$$

As shown in Chapter 17 and in Fig. 9, two magnetic elementary charges in the form of a north pole and a south pole are components of the proton.

This leads to the fundamental symmetry of an equal number of electric and magnetic elementary charges in the universe. This symmetry has been sought for a very long time.

The search for a unification of the fundamental interactions of nature is sometimes referred to as the search for the Holy Grail of physics. Generally, four fundamental interactions are considered. The interactions are described in numerous publications in connection with their respective exchange particles and their ranges of action. This will not be discussed in detail here. Instead, the focus will primarily be on the relationships between the interactions.

A dimensionless coupling constant is used as a measure of the interaction's coupling strength.

This uniform measure allows the interactions to be compared with one another. At high energies in the range of $200\ GeV$, the electromagnetic and weak interactions could be unified into the electroweak interaction.[54] The weak interaction will not be discussed further here.

Before examining the coupling constants of the other three interactions, I will discuss the underlying forces.

The gravitational force between an electron and a proton

$$F_{grav} = -G\frac{m_e\, m_p}{{r_{ep}}^2} \qquad \text{with} \qquad r_{ep} = \frac{m_e+m_p}{m_e\, m_p}\frac{\hbar}{\alpha_{em}\, c} \tag{310}$$

As an attractive force, the gravitational force is negative.

In equation (310), $\hbar$ is the reduced Planck constant and α_{em} is the well-known (electromagnetic) fine-structure constant. Since the masses of the electron m_e and the proton m_p are very small, the force F_{grav} is also very small.

[54] Brian Greene, The Elegant Universe, Siedler Verlag Berlin, 2000, ISBN 3-88680-699-5, p. 151

The distance r_{ep} between the two masses is $5.295\ 10^{-11}\ m$. This corresponds only approximately to the Bohr radius, since the calculation of r_{ep} includes both masses.[55]

A quantum mechanical analysis using the well-known radial wave function of the Schrödinger equation in spherical coordinates can show for the distance r_{ep}: The probability density of measuring the electron in the ground state of hydrogen reaches its maximum not at the Bohr radius, but at the radius r_{ep}. The gravitational force according to (310) with the radius r_{ep} is $F_{grav} = -3.628\ 10^{-47}\ N$.

The electric force between elementary charges

The well-known electric force between two elementary electric charges takes on a fixed value at a distance r_{ep}:

$$F_{el} = \frac{e^2}{4\pi\ \varepsilon_0\ r_{ep}{}^2} \tag{311}$$

It is $F_{el} = 8.230\ 10^{-8}\ N$.

The magnetic force between two elementary magnetic charges

It has been shown that the force between two elementary electric charges must be exactly equal to the force between two elementary magnetic charges when one charge is $p = 6.036\ 10^{-17}\ Vs$. However, magnetic monopoles have not yet been found.

$$F_{mg} = \frac{p^2}{4\pi\ \mu_0\ r_{ep}{}^2} \tag{312}$$

The force between an electric and a magnetic elementary charge

From equations (311) and (312), an electromagnetic force between an electric and a magnetic elementary charge can be derived:

$$F_{em} = \frac{ep}{4\pi\ \sqrt{\varepsilon_0}\sqrt{\mu_0}\ r_{ep}{}^2} = \frac{e\ p\ c}{4\pi\ \ r_{ep}{}^2} \tag{313}$$

At a distance of r_{ep} , it is also $8.230\ 10^{-8}\ N$.

The strong force between two quarks

For the strong force, formula (314) was derived from equation (293):

$$F_{strong} = \frac{\alpha_{strong}\ \hbar\ c}{r_{ep}{}^2} = \frac{\hbar\ c}{2\ r_{ep}{}^2} \tag{314}$$

[55] Andreas Gimsa, The Unification of the Fundamental Interactions, International Journal of Scientific Research and Management, Volume 09, Issue 05, May 2021, DOI: 10.18535/ijsrm/v9i05.p01, p.66

The strong force between two quarks, which theoretically have a distance of r_{ep}, is $F_{strong} = 5.639\ 10^{-6}\ N$ according to equation (314).

The ratios of the three fundamental forces to their respective coupling constants are equal:

$$\frac{F_{grav}}{\alpha_{grav}} = \frac{F_{em}}{\alpha_{em}} = \frac{F_{strong}}{\alpha_{strong}} = \frac{\hbar\, c}{{r_{ep}}^2} \tag{315}$$

The interaction constants

The known dimensionless coupling constant of the gravitational interaction α_{grav} can be calculated as follows:

$$\alpha_{grav} = \frac{\mathrm{G}\, m_e\, m_p}{\hbar\, c} = 3.217\ 10^{-42} \tag{316}$$

Gravity is the weakest of the three interactions. The coupling constant according to (316), which is derived here by taking into account the masses of the electron and proton, has a very low value. Although the proton mass and the electron mass decrease with the age of the universe, the coupling constant of the gravitational interaction remains constant.

The coupling constant of the electromagnetic interaction

The well-known dimensionless coupling constant of the electromagnetic interaction, α_{em} , can be calculated using the following formulas:

$$\alpha_{em} = \frac{e^2}{4\pi\, \varepsilon_0\, \hbar\, c} = \frac{p^2}{4\pi\, \mu_0\, \hbar\, c} = \frac{ep}{4\pi\, \sqrt{\varepsilon_0}\sqrt{\mu_0}\, \hbar\, c} = \frac{ep}{4\pi\, \hbar} = 7.297\ 10^{-3} \tag{317}$$

This coupling constant is the best-studied and has been known for a long time. It is sometimes also given as the ratio of the electron velocity in the hydrogen atom to the speed of light. This is approximately correct, but not exact. Rather, α_{em} is the ratio of the total velocity v_g of the electron and proton to the speed of light with the common period T: [56]

$$\alpha_{em} = \frac{2\pi\, r_{ep}}{Tc} = \frac{v_g}{c} \tag{318}$$

The coupling constant of the strong interaction

Equation (293) is briefly discussed here again in the direct context of the interactions:

$$\alpha_{strong} = \sqrt{\frac{\hbar}{m_p\, r_p\, c}} = 0.5$$

[56] Andreas Gimsa, The Metric of Spacetime, 2nd edition, Book on Demand GmbH, Hamburg, 2020, ISBN 978-3-00-064784-0, p. 89

At low energies and a quark separation of about $0.2\,fm$, the strong force leads to strong coupling of the quarks. The proton radius itself is approximately $0.8\,fm$. This coupling constant reaches its maximum value of $\alpha_{strong}(max) = 0.5$. At larger separations, the coupling constant no longer increases.

At higher energies and smaller distances, the coupling constant decreases, and the quarks then behave like free particles.[57]

However, the author believes that the strong force of the charges arises from the increased material-dependent permeability μ_r according to (200) and the material-dependent permittivity ε_r within the proton. This applies to the two magnetic elementary charges and the positron in the proton.

According to the principles of quantum physics, there is even a small probability of finding the hydrogen electron in the proton, i.e., in the vicinity of the positron. Gravity, which was derived in Chapter 18 from the expansion of the universe and thus of the atoms, results from electric and magnetic displacement currents. The calculation of the gravitational force includes the permeability μ_r and the permittivity ε_r present in the atoms. Both material-dependent quantities are greater in the proton than in the "empty" atom and in a vacuum.

In the proton, according to Equation (204), $\mu_r = \varepsilon_r = R_{Kl}/Z_0$ and thus $\mu_r = \varepsilon_r = F_{strong}/F_{em}$. Compared to the vacuum, the permeability and permittivity in the proton are therefore increased by a factor of 68.5. The following applies in Proton: $\sqrt{\mu_r \varepsilon_r} = F_{strong}/F_{em}$.

As the universe ages, the mass of the proton decreases and its radius increases. However, the product of mass and radius remains constant. Thus, the coupling constant of the strong interaction also remains constant.

A remarkable relationship between two interactions [58]

$$\alpha_{strong}\, \alpha_{grav}{}^2 = 0.5\, \alpha_{grav}{}^2 = \frac{m_e}{m_{uni}} \qquad (319)$$

Here we see yet another striking connection between the smallest particle, the electron, and the macroscopic system, the universe. Equations (202) and (292) yield yet another such relationship:

$$\frac{\alpha_{em}}{\alpha_{grav}} = \frac{F_{em}}{F_{grav}} = \frac{Z_0}{R_{Kl}}\, \frac{r_{uni}}{r_p} \qquad (319a)$$

---

[57] D.J. Gross and F. Wilczek, Asymptotically Free Gauge Theories, I, Phys. Rev. D8, 3633, 1973; Politzer, W., Asymptotic Freedom, An Approach to Strong Interactions, Phys. Rep. 14, 129, 1974
[58] Andreas Gimsa, The Unification of the Fundamental Interactions, International Journal of Scientific Research and Management, Volume 09, Issue 05, May 2021, DOI: 10.18535/ijsrm/v9i05.p01, p.69

21. Mathematical Connections between Quanta and the Universe

It should be noted once again at this point that the macroscopic and quantum physical perspectives differ from one another. For example, the proton radius should be regarded as the root mean square of the charge distribution. It is therefore "only" a typical or probable radius. This makes it all the more astonishing that the mathematical structure of nature apparently bridges this gap.

The following equivalent for the number of address Nats N_{uni} in the universe was found by the author through trial and error:

$$N_{uni} = \frac{m_{uni}\, t_{uni}\, c^2}{\hbar} = \sqrt{\frac{2\, m_{uni}^{\ 3}}{m_e\, m_p^{\ 2}}} = 6.546\ 10^{121} \tag{320}$$

Using equations (294) and (320), along with the mass of the universe $m_{uni} = t_{uni}\, c^3/G$, this new formula for the proton radius is obtained:

$$r_p = \frac{c\, t_{uni}^{\ 2}\, G^2\, m_e^{\ 2}\, m_p^{\ 3}}{\hbar^3} = 8.412\ 10^{-16}\, m \tag{321}$$

If the reader doubts the correctness of the following formulas, I suggest first checking whether the correct SI unit of the physical quantity being calculated is obtained. Afterward, the reader should verify the calculated quantity as such. If everything checks out so far, the reader will be amazed that the age of the universe is contained within the proton radius, a quantity supposedly fixed. Upon objective examination, the result is clear: the calculated quantities are time-dependent (see also Equation (135)).

For reasons of symmetry, the following formula could also apply to the electron radius:

$$r_e = \frac{c\, t_{uni}^{\ 2}\, G^2\, m_e^{\ 3}\, m_p^{\ 2}}{\hbar^3} = \frac{m_e\, r_p}{m_p} = 4.582\ 10^{-19}\, m \tag{322}$$

Although both the proton radius and the electron radius can be calculated very precisely using these formulas, formula (322) for the electron radius is speculative. The electron is often attributed a point-like structure, or it is described quantum mechanically only via its wave function.

Furthermore, the following would then apply to the electron radius:

$$r_e = \frac{2\, t_{uni}\, G\, m_e^{\ 2}}{\hbar} \tag{323}$$

With the proton radius r_p, the following interesting relationship holds with the electric and magnetic elementary charges e and p as well as the fine-structure constant α_{em}, *which is published here for the first time*:

$$\boldsymbol{r_p m_p c\,\pi\,\alpha_{em} = ep} \tag{324}$$

Using equations (294) and (321), the proton radius can be written as:

$$r_p = \frac{2\,t_{uni}\,G\,m_e\,m_p}{\hbar} \tag{325}$$

Here, too, it is easy to verify that the correct known value of the proton radius is obtained. Anyone who still doubts that the proton radius changes with the age of the universe should also explain here why Equation (325) yields the correct result with the variable age of the universe. The variability of G was already explained in Chapter 13.

The next equation, which can be derived using (135) and (294), shows a further relationship for calculating the proton radius r_p:

$$r_p = 2\sqrt{\frac{2\,t_{uni}\,G\,m_e}{c}} \tag{326}$$

Connection between Quanta and Gravitation

For the numbers N_1 and N_2 of the geometrically averaged charge masses according to Equation (274), Equation (281) applies in the two-mass system to both the electric and magnetic charges.

From this, the connection between magnetic elementary charges and Newton's gravitational force is to be established:

$$F_{grav} = G\frac{m_1\,m_2}{r^2} = \frac{N_1 N_2\,p^2}{4\pi\mu_0 r^2}\,\frac{\alpha_{grav}}{\alpha_{em}} \tag{327}$$

If one calculates this, for example, for the two-mass system Earth-Moon at their average distance, one arrives exactly at the well-known gravitational force of $F_{grav} = 1.985\;10^{20}\;N$. It must therefore hold that:

$$G = \frac{p^2}{4\pi\mu_0}\frac{1}{m_e\,m_p}\,\frac{\alpha_{grav}}{\alpha_{em}} \tag{328}$$

If we substitute the interaction constant from Equation (275) here, Equation (327) is confirmed. However, if one wishes to calculate the gravitational constant G, this can be done immediately using all known quantities, such as the age of the universe according to Equation (321). To avoid circular reasoning, the proton radius in (321) should be calculated according to (294). The following formula can also be used directly: [59]

[59] Andreas Gimsa, The Unification of the Fundamental Interactions, International Journal of Scientific Research and Management, Volume 09, Issue 05, May 2021, DOI: 10.18535/ijsrm/v9i05.p01, p.69

$$G = \frac{2\hbar^2}{c\, t_{uni}\, m_e m_p{}^2} \tag{329}$$

Further manipulations of equations (204), (325), and (328) lead to novel formulas for the Newtonian gravitational constant G:

$$G = \frac{1}{m_e m_p} \frac{r_p}{r_{uni}} \frac{R_{Kl}}{Z_0} \frac{e^2}{4\pi\varepsilon_0} = \frac{1}{m_e m_p} \frac{r_p}{r_{uni}} \frac{R_{Kl}}{Z_0} \frac{p^2}{4\pi\mu_0} = \frac{1}{m_e m_p} \frac{r_p}{r_{uni}} \frac{R_{Kl}}{Z_0} \frac{epc}{4\pi} \tag{330}$$

In my view, it is not surprising that the variable gravitational constant G incorporates the age of the universe or its radius. After all, gravity arises from the extremely slow expansion of atoms during the expansion of the universe (Chapter 18). Even quantum particles cannot escape global temporality.

The correctness of Equation (330) can be easily verified. The gravitational constant is thus, as is known: $G = 6.674\ 10^{-11}\ m^3/(kg\ s^2)$. This also makes it clear that the gravitational constant must change with the age of the universe. For a reduction in the mass of electrons and protons occurs over time.

Using equations (204), (329), (330), and the known relationship $e^2 = 4\pi\ \alpha_{em} c\ \varepsilon_0\ \hbar$, the following calculation formula for the proton radius follows:

$$r_p = \frac{\hbar}{\alpha_{em}{}^2\, c\ m_p} \frac{Z_0{}^2}{R_{Kl}{}^2} \tag{331}$$

Thus, the time-varying proton mass m_p alone determines the time-varying proton radius. All other quantities in equation (331) are true, unchanging natural constants.

Alan Guth describes that a massive, compressible spherical shell, which contracts under its own gravity, generates an external gravitational field that was not previously present in the same region of space within the spherical shell. He deduced from this that energy is released in the process and that the gravitational energy must therefore be negative. [60]

In principle, a reverse process could occur during the expansion of the universe. The universe's external gravitational field is dissipated, and energy is required for this. The energy could come from the masses. These would have to decrease with the age of the universe. This would also explain why, for the universe, the magnitude of the mass energy corresponds exactly to that of the gravitational energy (see Equation (65)).

---

[60] Alan Guth, The Birth of the Cosmos from Nothing, Droemersche Verlagsanstalt Munich, 2002, ISBN 3-426-77610-3, p.466

In addition to the fascinating Equation (135), further new formulas for the age of the universe can be derived from Equations (320) through (330):

$$t_{uni} = \sqrt{\frac{2\, m_{uni}}{m_e} \frac{\hbar}{m_p c}} \tag{332}$$

$$t_{uni} = \frac{m_{uni}}{m_p\, m_e} \frac{e\, p}{4\pi\, c^2} \frac{r_p}{r_{uni}} \frac{R_{Kl}}{Z_0} = \sqrt{\frac{m_{uni}}{4\pi} \frac{R_{Kl}}{Z_0} \frac{r_p}{c^3} \frac{ep}{m_e m_p}} \tag{333}$$

These two formulas also demonstrate the direct connection between the universe—in terms of its age and mass—and the smallest constituents of matter, namely protons.

Concluding Remarks

"Mathematics is the instrument that mediates between theory and practice, between thought and observation: it builds the connecting bridge and makes it ever more robust. This is why our entire contemporary culture, insofar as it is based on the intellectual penetration and harnessing of nature, finds its foundation in mathematics." [61]

The universe has been explained as an information-processing system whose temporal loss of mass is used to provide the energy necessary for storing information. More and more of the existing constant potential information in the form of microstates becomes addressable. This can create the impression of constantly growing entropy, yet entropy remains constant in the universe for all eternity.

The origin of time has been described, and it was recognized that time has an information equivalent, just as every mass has an energy equivalent according to Einstein.

The universe has been described as a pulsating system that does not necessarily continue to expand indefinitely and die a heat death.

It has been demonstrated that, over time, important physical quantities previously considered constant must change. These include all object masses and Newton's gravitational constant.

The constant increase in all distances was demonstrated, for example, using the Earth-Moon distance.

A plausible and simple structure of the proton was found, resulting in charge symmetry in the universe.

Gravitational force has been described as a Lorentz force acting between the electric and magnetic charges of atoms.

Even if assumptions turn out to be implausible or incorrect, they are still capable of advancing the truth. It is always the combination of time and ideas that ultimately brings light into the darkness and makes the world explainable. The natural principle of trial and error also applies in physics.

I hope that the new evidence and ideas I have presented to the reader in this book will be highly useful. May they fall on fertile ground and be further developed.

---

[61] David Hilbert, The Understanding of Nature and Logic, Königsberg, 1930; Assembly of German Natural Scientists and Physicians; Collected Papers, Volume Three, Published by Julius Springer, Berlin, 1935, p. 385